The Scientist as Rebel

The Scientist as Rebel

Freeman Dyson

NEW YORK REVIEW BOOKS

New York

THIS IS A NEW YORK REVIEW BOOK

PUBLISHED BY THE NEW YORK REVIEW OF BOOKS

THE SCIENTIST AS REBEL
by Freeman Dyson

This edition published in 2008
in the United States of America by
The New York Review of Books
435 Hudson Street
New York, NY 10014
www.nyrb.com

The Library of Congress has catalogued the hardcover edition of this book as follows:

Dyson, Freeman J.
The Scientist as Rebel / by Freeman J. Dyson.
p. cm.
Includes bibliographical references.
ISBN-13: 978-1-59017-216-2 (alk. paper)
ISBN-10: 1-59017-216-7 (alk. paper)
1. Science. 2. Physics. 3. Science—History. 4. Science—Moral and ethical aspects.
5. Dyson, Freeman J.—Biography. 6. Scientists—United States—Biography. I. Title.
Q158.5.D977 2006
500—dc22

2006022081

ISBN 978-1-59017-294-0

Printed in the United States of America on acid-free paper.

1 3 5 7 9 10 8 6 4 2

To my teachers,
Eric and Cordelia James

Great brow, frail frame—gone. Yet you abide
In the shadow and sheen,
All the mellowing traits of a countryside
That nursed your tragi-comical scene;
And in us, warmer-hearted and brisker-eyed
Since you have been.

—Cecil Day Lewis

Table of Contents

Preface

BENJAMIN FRANKLIN COMBINED better than anyone else the qualities of a great scientist and a great rebel. As a scientist, without formal education or inherited wealth, he beat the learned aristocrats of Europe at their own game. His victory encouraged him to believe that he and his fellow citizens in America, without much training in military strategy or international politics, could beat the aristocrats of Europe at warfare and diplomacy. Franklin's triumph as a rebel resulted from the fact that his rebellion was not impulsive but was carefully thought out over many years. For most of his long life, he was a loyal subject of the British King. He lived for many years in London, representing the Commonwealth of Pennsylvania in dealings with the British government, calmly taking the measure of his future enemies.

While he was in London, Franklin was an active member of the Society for the Encouragement of Arts, Manufactures and Commerce, which still flourishes today. The society encouraged inventions and manufactures by offering financial subsidies and prizes to inventors and entrepreneurs. The prizes were usually available to all subjects of the King in England or America, but they were often targeted to subsidize colonial enterprises that the society considered desirable. When Franklin first joined the society in 1755, he was an enthusiastic supporter of its efforts to encourage invention, which he saw as complementary to the efforts of his own Philosophical Society in America. But as the years went by, his attitude became more critical. He never openly disagreed with the society and remained a member in good standing, all through the War of Independence and afterward until his death. But he recorded privately, in the margin of a book, his true feelings about the system of prizes and subsidies offered by the society:

> What you call Bounties given by Parliament and the Society are nothing more than Inducements offered to us, to induce us to leave Employments that are more profitable and engage in such as would be less so without your Bounty; to quit a Business profitable to ourselves and engage in one as shall be profitable to you; this is the true Spirit of all your Bounties.

He wrote these words in 1770, five years before the outbreak of the war that ended British rule in the thirteen colonies.

Franklin became a rebel only when he judged the time to be ripe and the costs to be acceptable. As a rebel he remained a conservative, aiming not to destroy but to preserve as much as possible of the established order of society. As a diplomat in Paris, he fitted smoothly into the established order of prerevolutionary France. He would not have fitted so well into the France of Danton and Robespierre ten years later. The rebellion that Franklin embodied was a thoughtful rebellion, driven by reason and calculation more than by passion and hatred.

In spite of its title, this book is mostly not about rebel scientists. It is a collection of book reviews, prefaces, and essays on a variety of subjects. The majority were published in *The New York Review of Books*. I am grateful to *The New York Review* for inviting me to collect these in a book, and for allowing me to supplement them with other pieces that were published in other places. The bibliographical notes at the end explain where each piece was published and how it originated.

The collection is divided into four sections according to subject matter, and arranged chronologically within each section. Section I deals with political issues arising out of science and technology. Section II deals with problems of war and peace. Section III deals with the history of science, and Section IV with personal and philosophical reflections. By accident rather than by design, at least one rebellious scientist appears in each section. But there are pieces about scientists such as John Cockcroft and Ernest Walton (Chapter 21) who were far from being rebels, and pieces such as the review of Max Hastings's *Armageddon* (Chapter 13) that are concerned with soldiers rather than with scientists.

One of the pleasures of writing for *The New York Review* is the fact that it publishes long reviews. The reviewer is asked to write about four thousand words, which means that the review can be an essay reflecting on the subject matter rather than a simple appraisal of a book. The short reviews in this collection were published in other journals. If this book is a sandwich, the meat is the series of twelve long reviews from *The New York Review*, most of them appearing in Section III. There are four other meaty items that were not in *The New York Review*. One is the Bernal lecture (Chapter 24), which Carl Sagan whimsically published as an appendix to the proceedings of a conference on communication with extraterrestrial intelligence. The other three (Chapters 8, 9, and 10) are chapters from my book *Weapons and Hope*, which is now out of print. The collapse of the Soviet Union made much of *Weapons and Hope* obsolete, but these three historical chapters may be worth preserving.

The essay "The Scientist as Rebel," with which this collection begins, originated as a talk given at a meeting of scientists and philosophers at Cambridge, England, in November 1992. The talk was dedicated to the memory of Lord James of Rusholme, who had died six months earlier at the age of eighty-three, full of years and honors, having risen to the top of the British educational establishment. The obituary notices that were published in newspapers after his death described him as a capable organizer and administrator who presided over the founding of York University and served as its vice-chancellor for the first eleven years of its existence, from 1962 to 1973. They said that he had conservative views on the subject of education, that he believed in old-fashioned scholarship and academic rigor, that he fought hard to make York University a community of scholars and an intellectual powerhouse on a level with Oxford. "Jude the Obscure," he was quoted as saying, "need no longer look despairingly at the towers and spires of an inaccessible university, provided he has three good A-level passes, can satisfy one of a multiplicity of entrance requirements, and is prepared, if necessary, to do without spires." He tried to make York University the home of an intellectual elite, an elite based upon brains and competitive examinations rather than money and social class. His elitist view of education came into collision with the dominant political currents of the 1950s and 1960s. The dominant view held that Jude should be enrolled in a university whether or not he was able to pass the A-level examinations. The dominant view held that higher education should be for everybody and not only for the bright. In the end, Lord James fought in vain against what he considered the folly of the politicians. Whenever he lost a battle in his campaign for strict intellectual standards, he liked to quote the lines of the poet Matthew Arnold:

Let the victors, when they come,
When the forts of folly fall,
Find thy body by the wall!

I dedicated "The Scientist as Rebel" to Lord James because he was, like Benjamin Franklin, a scientist and a rebel. Like Franklin, he achieved great things as a rebel because he was aiming to build a new society rather than to destroy an old one. Like Franklin, he built institutions to last. After he had achieved his goal of building a new university, he was a conservative administrator. But I knew him very well thirty years earlier, long before any of us dreamed that he might one day be sitting in the House of Lords. In those days he was plain Eric James, a teacher of chemistry in the school at Winchester where I was a boy. He had published a successful textbook, *Elements of Physical Chemistry*, that was widely used in schools. He was indeed a scientist, and he was a rebel and an outsider, who brought a draft of fresh air into the stuffy old chambers of Winchester College. But he also understood the value of tradition. He was big enough to see both sides of the picture. At Winchester, where intellectual traditions are taken for granted, we saw Eric the reformer. At York in the 1960s, when intellectual standards were everywhere under attack, we saw Eric the traditionalist. Between Winchester and York he spent seventeen years as high master of Manchester grammar school. At Manchester in the postwar years he occupied the middle ground in a society rebuilding itself. Manchester gave him the opportunity to combine the two main purposes of his life, the education of gifted children and the reform of society.

My most vivid memory of Eric comes from the summer of 1941. Since many of the regular farmworkers had been drafted into the army, schoolchildren and teachers were invited to help out on the farms during school vacations. We were encamped together for two rain-drenched weeks at Hurstbourne Priors in rural Hampshire, trying to rescue a sodden harvest of wheat and oats, with the grain already sprouting green out of the sheaves. In those days the farmers did not have heated drying sheds. A wet August meant a spoiled harvest. We worked in the fields all day and discussed the meaning of existence in our tents at night. Those two weeks were in retrospect the high point of my school days, breaking out of the academic cocoon and seeing something of the world outside, with Brechtian commentaries provided by Eric and his wife, Cordelia. Cordelia fought bravely for fifty years at Eric's side against the forts of folly. At Hurstbourne

Priors Eric and Cordelia came into collision with Lord Lymington, who owned the land on which we were working. This was the same Lord Lymington who appears in Chapter 17 of this book, the review of James Gleick's biography of Newton. Lord Lymington had inherited Newton's manuscripts and carelessly dispersed them all over the world by selling them at auction in small lots. Eric and Cordelia entertained us at night with accurate imitations of Lord Lymington's high-pitched voice and fatuous oratory.

When Eric James died in 1992, the film *Dead Poets Society* was playing in movie theaters. It is a story about an upper-class American prep school and an English teacher who gets into trouble because he doesn't stick to the established curriculum. The theme of the film is rebellion. The established curriculum is asinine, the headmaster is a stuffed shirt, and the only redeeming feature of the school is the English teacher and a bunch of rebellious boys whom he encourages to break the rules. The film was a fitting memorial to Eric. Our school in Winchester was like the school in the film. The atmosphere was the same, with the rebellious boys and the smooth-talking headmaster. Instead of holding meetings in a cave at night, we took advantage of the wartime blackout to climb over the rooftops and up the chapel tower. And instead of a subversive English teacher we had our subversive chemistry teacher. Like the teacher in the film, Eric James had a passion for poetry. He had a Ph.D. in chemistry, but he understood that it made no sense to bore us with formal lectures about chemical reactions which we could learn about much quicker from textbooks. So he put aside the ferrous and ferric oxides and read us the latest poems of Auden and Isherwood and Dylan Thomas and Cecil Day Lewis, the poets who were then speaking for the younger generation in the first desperate years of World War II.

Forty years later I met Eric James at a party at York University, after his retirement as vice-chancellor. It was the first time I had seen him since I was seventeen. I started the conversation with a quote from one of the poems he had read to us forty years earlier, a poem by Day Lewis about the war in Spain:

They bore not a charmed life.
They went into battle foreseeing
Probable loss, and they lost.

Eric continued without a break from his own memory:

The tides of Biscay flow
Over the obstinate bones of many,
the winds are sighing
Round prison walls where the rest are
doomed like their ship to rust,
Men of the Basque country, the Mar Cantabrico.

Fortunately our headmaster, unlike the headmaster in the film, was wise enough to tolerate Eric James and give him a free hand. Eric was accepted into the English educational hierarchy, became a headmaster himself, founded a university, and was rewarded by a grateful government with a baronial title. It is hard to imagine a prep school chemistry teacher in the United States ending his career in so exalted a fashion. But Eric remained in his heart a rebel. Through forty years of active and creative life he remembered the sadness and the passion of the 1940s when we saw Hell break loose on Earth. That sadness and that passion are a part of our lives still. That sadness and that passion are what made Eric James a great teacher.

The life of Eric James demonstrates that there is no contradiction between a rebellious spirit and an uncompromising pursuit of excellence in a rigorous intellectual discipline. In the history of science, it has often happened that rebellion and professional competence went hand in hand. Several chapters in this book are devoted to famous scientists who were also famous rebels. Thomas Gold (Chapter 3) was a great astronomer with heretical opinions about many subjects. Joseph Rotblat (Chapter 12) was unique as a scientist who walked out of the wartime Los Alamos bomb project when he learned that the threat of a German atomic bomb had disappeared. Norbert Wiener (Chapter 22) was a great mathematician who refused on moral grounds to have anything to do with either industry or government. Desmond Bernal (Chapter 24) was one of the founding fathers of molecular biology, and also a faithful member of the Communist Party and a passionate believer in Marxism. Three chapters (23, 25, and 26) are devoted to my teacher Richard Feynman, the physicist who most closely resembled Eric James. Feynman was another rebellious spirit who combined a serious dedication to science with joyful adventures in the world outside.

The scientist who described most eloquently the role of the rebel in science was the paleontologist Loren Eiseley. Unfortunately Eiseley does not have a chapter in this book. He was a wonderful writer, best known to the general public through his books *The Immense Journey* and *The Unexpected Universe*, which tell poignant stories about the creatures, living and dead, that Eiseley encountered in the course of his work as a naturalist and fossil hunter. The most personal of his books is his autobiography, *All the Strange Hours*. In it Eiseley explains why he is a rebel, why he is a poet, why he feels less kinship with his academic colleagues than with a doomed prisoner escaped from jail on a winter's night and hunted to death in the snow. Eiseley's image of the prisoner bleeding in the snow, Day Lewis's image of the Spanish sailors rusting in Franco's prison, both are images of the human condition as valid today as they were sixty years ago.

—Freeman Dyson, Princeton, 2006

I

Contemporary Issues in Science

I

THE SCIENTIST AS REBEL

THERE IS NO such thing as a unique scientific vision, any more than there is a unique poetic vision. Science is a mosaic of partial and conflicting visions. But there is one common element in these visions. The common element is rebellion against the restrictions imposed by the locally prevailing culture, Western or Eastern as the case may be. The vision of science is not specifically Western. It is no more Western than it is Arab or Indian or Japanese or Chinese. Arabs and Indians and Japanese and Chinese had a big share in the development of modern science. And two thousand years earlier, the beginnings of ancient science were as much Babylonian and Egyptian as Greek. One of the central facts about science is that it pays no attention to East and West and North and South and black and yellow and white. It belongs to everybody who is willing to make the effort to learn it. And what is true of science is also true of poetry. Poetry was not invented by Westerners. India has poetry older than Homer. Poetry runs as deep in Arab and Japanese culture as it does in Russian and English. Just because I quote poems in English, it does not follow that the vision of poetry has to be Western. Poetry and science are gifts given to all of humanity.

For the great Arab mathematician and astronomer Omar Khayyam, science was a rebellion against the intellectual constraints of Islam, a rebellion which he expressed more directly in his incomparable verses:

And that inverted Bowl they call the Sky,
Whereunder crawling cooped we live and die,
Lift not your hands to It for help,
—for It
As impotently rolls as you or I.

For the first generations of Japanese scientists in the nineteenth century, science was a rebellion against their traditional culture of feudalism. For the great Indian physicists of this century, Raman, Bose, and Saha, science was a double rebellion, first against English domination and second against the fatalistic ethic of Hinduism. And in the West, too, great scientists from Galileo to Einstein have been rebels. Here is how Einstein himself described the situation:

> When I was in the seventh grade at the Luitpold Gymnasium in Munich, I was summoned by my home-room teacher who expressed the wish that I leave the school. To my remark that I had done nothing amiss, he replied only, "Your mere presence spoils the respect of the class for me."

Einstein was glad to be helpful to the teacher. He followed the teacher's advice and dropped out of school at the age of fifteen.

From these and many other examples we see that science is not governed by the rules of Western philosophy or Western methodology. Science is an alliance of free spirits in all cultures rebelling against the local tyranny that each culture imposes on its children. Insofar as I am a scientist, my vision of the universe is not reductionist or anti-reductionist. I have no use for Western isms of any kind. I feel myself a traveler on the "Immense Journey" of the paleontologist Loren

Eiseley, a journey that is far longer than the history of nations and philosophies, longer even than the history of our species.

A few years ago an exhibition of Paleolithic cave art came to the Museum of Natural History in New York. It was a wonderful opportunity to see in one place the carvings in stone and bone that are normally kept in a dozen separate museums in France. Most of the carvings were done in France about 14,000 years ago, during a short flowering of artistic creation at the very end of the last ice age. The beauty and delicacy of the carving is extraordinary. The people who carved these objects cannot have been ordinary hunters amusing themselves in front of the cave fire. They must have been trained artists sustained by a high culture.

And the greatest surprise, when you see these objects for the first time, is the fact that their culture is not Western. They have no resemblance at all to the primitive art that arose 10,000 years later in Mesopotamia and Egypt and Crete. If I had not known that the old cave art was found in France, I would have guessed that it came from Japan. The style looks today more Japanese than European. That exhibition showed us vividly that over periods of 10,000 years the distinctions between Western and Eastern and African cultures lose all meaning. Over a time span of 100,000 years we are all Africans. And over a time span of 300 million years we are all amphibians, waddling uncertainly out of dried-up ponds onto the alien and hostile land.

And with this long view of the past goes Robinson Jeffers's even longer view of the future. In the long view, not only European civilization but the human species itself is transitory. Here is the vision of Robinson Jeffers, expressed in different parts of his long poem "The Double Axe."

> *"Come, little ones.*
> *You are worth no more than the foxes and yellow*
> *wolfkins, yet I will give you wisdom.*

O future children:
Trouble is coming; the world as of the present time
Sails on its rocks; but you will be born and live
Afterwards. Also a day will come when the earth
Will scratch herself and smile and rub off humanity:
But you will be born before that."

"Time will come, no doubt,
When the sun too shall die; the planets will freeze,
and the air on them; frozen gases, white flakes of air
Will be the dust: which no wind ever will stir: this very
dust in dim starlight glistening
Is dead wind, the white corpse of wind.
Also the galaxy will die; the glitter of the Milky Way,
our universe, all the stars that have names are dead.
Vast is the night. How you have grown, dear night,
walking your empty halls, how tall!"[1]

Robinson Jeffers was no scientist, but he expressed better than any other poet the scientist's vision. Ironic, detached, contemptuous like Einstein of national pride and cultural taboos, he stood in awe of nature alone. He stood alone in uncompromising opposition to the follies of the Second World War. His poems during those years of patriotic frenzy were unpublishable. "The Double Axe" was finally published in 1948, after a long dispute between Jeffers and his editors. I discovered Jeffers thirty years later, when the sadness and the passion of the war had become a distant memory. Fortunately, his works are now in print and you can read them for yourselves.

Science as subversion has a long history. There is a long list of

1. Robinson Jeffers, *The Double Axe and Other Poems, including eleven suppressed poems* (Liveright, 1977).

scientists who sat in jail and of other scientists who helped get them out and incidentally saved their lives. In our century we have seen the physicist Lev Landau sitting in jail in the Soviet Union and Pyotr Kapitsa risking his own life by appealing to Stalin to let Landau out. We have seen the mathematician André Weil sitting in jail in Finland during the Winter War of 1939–1940 and Lars Ahlfors saving his life. The finest moment in the history of the Institute for Advanced Study, where I work, came in 1957, when we appointed the mathematician Chandler Davis a member of the institute, with financial support provided by the American government through the National Science Foundation. Davis was then a convicted felon because he refused to rat on his friends when questioned by the House Un-American Activities Committee. He had been convicted of contempt of Congress for not answering questions and had appealed his conviction to the Supreme Court.

While his case was under appeal, he came to Princeton and continued doing mathematics. That is a good example of science as subversion. After his institute fellowship was over, he lost his appeal and sat for six months in jail. Davis is now a distinguished professor at the University of Toronto and is actively engaged in helping people in jail to get out. Another example of science as subversion is Andrei Sakharov. Davis and Sakharov belong to an old tradition in science that goes all the way back to the rebels Benjamin Franklin and Joseph Priestley in the eighteenth century, to Galileo and Giordano Bruno in the seventeenth and sixteenth. If science ceases to be a rebellion against authority, then it does not deserve the talents of our brightest children. I was lucky to be introduced to science at school as a subversive activity of the younger boys. We organized a Science Society as an act of rebellion against compulsory Latin and compulsory football. We should try to introduce our children to science today as a rebellion against poverty and ugliness and militarism and economic injustice.

The vision of science as rebellion was articulated in Cambridge with great clarity on February 4, 1923, in a lecture by the biologist J. B. S. Haldane to the Society of Heretics. The lecture was published as a little book with the title *Daedalus*. Here is Haldane's vision of the role of scientist. I have taken the liberty to abbreviate Haldane slightly and to omit the phrases that he quoted in Latin and Greek, since unfortunately I can no longer assume that the heretics of Cambridge are fluent in those languages.

> The conservative has but little to fear from the man whose reason is the servant of his passions, but let him beware of him in whom reason has become the greatest and most terrible of the passions. These are the wreckers of outworn empires and civilizations, doubters, disintegrators, deicides. In the past they have been men like Voltaire, Bentham, Thales, Marx, but I think that Darwin furnishes an example of the same relentlessness of reason in the field of science. I suspect that as it becomes clear that at present reason not only has a freer play in science than elsewhere, but can produce as great effects on the world through science as through politics, philosophy or literature, there will be more Darwins.
>
> We must regard science, then, from three points of view. First, it is the free activity of man's divine faculties of reason and imagination. Secondly, it is the answer of the few to the demands of the many for wealth, comfort and victory, gifts which it will grant only in exchange for peace, security and stagnation. Finally it is man's gradual conquest, first of space and time, then of matter as such, then of his own body and those of other living beings, and finally the subjugation of the dark and evil elements in his own soul.[2]

2. J. B. S. Haldane, *Daedalus, or Science and the Future* (London: Kegan Paul, 1924).

I have already made it clear that I have a low opinion of reductionism, which seems to me to be at best irrelevant and at worst misleading as a description of what science is about. Let me begin with pure mathematics. Here the failure of reductionism has been demonstrated by rigorous proof. This will be a familiar story to many of you. The great mathematician David Hilbert, after thirty years of high creative achievement on the frontiers of mathematics, walked into a blind alley of reductionism. In his later years he espoused a program of formalization, which aimed to reduce the whole of mathematics to a collection of formal statements using a finite alphabet of symbols and a finite set of axioms and rules of inference. This was reductionism in the most literal sense, reducing mathematics to a set of marks written on paper, and deliberately ignoring the context of ideas and applications that give meaning to the marks. Hilbert then proposed to solve the problems of mathematics by finding a general process that could decide, given any formal statement composed of mathematical symbols, whether that statement was true or false. He called the problem of finding this decision process the *Entscheidungsproblem*. He dreamed of solving the *Entscheidungsproblem* and thereby solving as corollaries all the famous unsolved problems of mathematics. This was to be the crowning achievement of his life, the achievement that would outshine all the achievements of earlier mathematicians who solved problems only one at a time.

The essence of Hilbert's program was to find a decision process that would operate on symbols in a purely mechanical fashion, without requiring any understanding of their meaning. Since mathematics was reduced to a collection of marks on paper, the decision process should concern itself only with the marks and not with the fallible human intuitions out of which the marks were reduced. In spite of the prolonged efforts of Hilbert and his disciples, the *Entscheidungsproblem* was never solved. Success was achieved only in highly restricted domains of mathematics, excluding all the deeper and more interesting concepts. Hilbert never gave up hope, but as the years went by his program became an exercise in formal logic having little connection with real mathematics. Finally, when Hilbert was seventy years old, Kurt Gödel proved by a brilliant analysis that the *Entscheidungsproblem* as Hilbert formulated it cannot be solved.

Gödel proved that in any formulation of mathematics, including the rules of ordinary arithmetic, a formal process for separating statements into true and false cannot exist. He proved the stronger result which is now known as Gödel's theorem, that in any formalization of mathematics including the rules of ordinary arithmetic there are meaningful arithmetical statements that cannot be proved true or false. Gödel's theorem shows conclusively that in pure mathematics reductionism does not work. To decide whether a mathematical statement is true, it is not sufficient to reduce the statement to marks on paper and to study the behavior of the marks. Except in trivial cases, you can decide the truth of a statement only by studying its meaning and its context in the larger world of mathematical ideas.

It is a curious paradox that several of the greatest and most creative spirits in science, after achieving important discoveries by following their unfettered imaginations, were in their later years obsessed with reductionist philosophy and as a result became sterile. Hilbert was a prime example of this paradox. Einstein was another. Like Hilbert, Einstein did his great work up to the age of forty without any reductionist bias. His crowning achievement, the general relativistic theory of gravitation, grew out of a deep physical understanding of natural processes. Only at the very end of his ten-year struggle to understand gravitation did he reduce the outcome of his understanding to a finite set of field equations. But like Hilbert, as he grew older he concentrated his attention more and more on the formal properties of his equations, and he lost interest in the wider universe of ideas out of which the equations arose.

His last twenty years were spent in a fruitless search for a set of equations that would unify the whole of physics, without paying attention to the rapidly proliferating experimental discoveries that any unified theory would finally have to explain. I do not need to say more about this tragic and well-known story of Einstein's lonely attempt to reduce physics to a finite set of marks on paper. His attempt failed as dismally as Hilbert's attempt to do the same thing with mathematics. I shall instead discuss another aspect of Einstein's later life, an aspect that has received less attention than his quest for the unified field equations: his extraordinary hostility to the idea of black holes.

Black holes were invented by J. Robert Oppenheimer and Hartland Snyder in 1939. Starting from Einstein's theory of general relativity, Oppenheimer and Snyder found solutions of Einstein's equations that described what happens to a massive star when it has exhausted its supplies of nuclear energy. The star collapses gravitationally and disappears from the visible universe, leaving behind only an intense gravitational field to mark its presence. The star remains in a state of permanent free fall, collapsing endlessly inward into the gravitational pit without ever reaching the bottom. This solution of Einstein's equations was profoundly novel. It has had enormous impact on the later development of astrophysics.

We now know that black holes ranging in mass from a few suns to a few billion suns actually exist and play a dominant role in the economy of the universe. In my opinion, the black hole is incomparably the most exciting and the most important consequence of general relativity. Black holes are the places in the universe where general relativity is decisive. But Einstein never acknowledged his brainchild. Einstein was not merely skeptical, he was actively hostile to the idea of black holes. He thought that the black hole solution was a blemish to be removed from his theory by a better mathematical formulation, not a consequence to be tested by observation. He never expressed the slightest enthusiasm for black holes, either as a concept or as a physical possibility. Oddly enough, Oppenheimer too in later life was uninterested in black holes, although in retrospect we can say that they were his most important contribution to science. The older Einstein and the older Oppenheimer were blind to the mathematical beauty of black holes, and indifferent to the question whether black holes actually exist.

How did this blindness and this indifference come about? I never discussed this question directly with Einstein, but I discussed it several times with Oppenheimer and I believe that Oppenheimer's answer applies equally to Einstein. Oppenheimer in his later years believed that the only problem worthy of the attention of a serious theoretical physicist was the discovery of the fundamental equations of physics. Einstein certainly felt the same way. To discover the right equations was all that mattered. Once you had discovered the right equations, then the study of particular solutions of the equations would be a routine exercise for second-rate physicists or graduate students. In Oppenheimer's view, it would be a waste of his precious time, or of mine, to concern ourselves with the details of particular solutions. This was how the philosophy of reductionism led Oppenheimer and Einstein astray. Since the only purpose of physics was to reduce the world of physical phenomena to a finite set of fundamental equations, the study of particular solutions such as black holes was an undesirable distraction from the general goal. Like Hilbert, they were not content to solve particular problems one at a time. They were entranced by the dream of solving all the basic problems at once. And as a result, they failed in their later years to solve any problems at all.

In the history of science it happens not infrequently that a reductionist approach leads to a spectacular success. Frequently the understanding of a complicated system as a whole is impossible without an understanding of its component parts. And sometimes the understanding of a whole field of science is suddenly advanced by the discovery of a single basic equation. Thus it happened that the Schrödinger equation in 1926 and the Dirac equation in 1927 brought a miraculous order into the previously mysterious processes of atomic physics. The equations of Erwin Schrödinger and Paul Dirac were triumphs of reductionism. Bewildering complexities of chemistry and physics were reduced to two lines of algebraic symbols. These triumphs were in Oppenheimer's mind when he belittled his own discovery of black holes. Compared with the abstract beauty and simplicity of the Dirac equation, the black hole solution seemed to him ugly, complicated, and lacking in fundamental significance.

But it happens at least equally often in the history of science that the understanding of the component parts of a composite system is impossible without an understanding of the behavior of the system as a whole. And it often happens that the understanding of the mathematical nature of an equation is impossible without a detailed understanding of its solutions. The black hole is a case in point. One could say without exaggeration that Einstein's equations of general relativity were understood only at a very superficial level before the discovery of the black hole. During the fifty years since the black hole was invented, a deep mathematical understanding of the geometrical structure of space-time has slowly emerged, with the black hole solution playing a fundamental role in the structure. The progress of science requires the growth of understanding in both directions, downward from the whole to the parts and upward from the parts to the whole. A reductionist philosophy, arbitrarily proclaiming that the growth of understanding must go only in one direction, makes no scientific sense. Indeed, dogmatic philosophical beliefs of any kind have no place in science.

Science in its everyday practice is much closer to art than to philosophy. When I look at Gödel's proof of his undecidability theorem, I do not see a philosophical argument. The proof is a soaring piece of architecture, as unique and as lovely as Chartres Cathedral. Gödel took Hilbert's formalized axioms of mathematics as his building blocks and built out of them a lofty structure of ideas into which he could finally insert his undecidable arithmetical statement as the keystone of the arch. The proof is a great work of art. It is a construction, not a reduction. It destroyed Hilbert's dream of reducing all mathematics to a few equations, and replaced it with a greater dream of mathematics as an endlessly growing realm of ideas. Gödel proved that in mathematics the whole is always greater than the sum of the parts. Every formalization of mathematics raises questions that reach beyond the limits of the formalism into unexplored territory.

The black hole solution of Einstein's equations is also a work of art. The black hole is not as majestic as Gödel's proof, but it has the essential features of a work of art: uniqueness, beauty, and unexpectedness. Oppenheimer and Snyder built out of Einstein's equations a structure that Einstein had never imagined. The idea of matter in permanent free fall was hidden in the equations, but nobody saw it until it was revealed in the Oppenheimer-Snyder solution. On a much more humble level, my own activities as a theoretical physicist have a similar quality. When I am working, I feel myself to be practicing a craft rather than following a method. When I did my most important piece of work as a young man, putting together the ideas of Sin-Itiro Tomonaga, Julian Schwinger, and Richard Feynman to obtain a simplified version of quantum electrodynamics, I had consciously in mind a metaphor to describe what I was doing. The metaphor was bridge-building. Tomonaga and Schwinger had built solid foundations on one side of a river of ignorance, Feynman had built solid foundations on the other side, and my job was to design and build the cantilevers reaching out over the water until they met in the middle. The metaphor was a good one. The bridge that I built is still serviceable and still carrying traffic forty years later. The same metaphor describes well the greater work of unification achieved by Stephen Weinberg and Abdus Salam when they bridged the gap between electrodynamics and the weak interactions. In each case, after the work of unification is done, the whole stands higher than the parts.

In recent years there has been great dispute among historians of science, some believing that science is driven by social forces, others believing that science transcends social forces and is driven by its own internal logic and by the objective facts of nature. Historians of the first group write social history, those of the second group write intellectual history. Since I believe that scientists should be artists and rebels, obeying their own instincts rather than social demands or philosophical principles, I do not fully agree with either view of history. Nevertheless, scientists should pay attention to the historians. We have much to learn, especially from the social historians.

Many years ago, when I was in Zürich, I went to see the play *The Physicists* by the Swiss playwright Friedrich Dürrenmatt. The characters in the play are grotesque caricatures, wearing the costumes and using the names of Newton, Einstein, and Möbius. The action takes place in a lunatic asylum where the physicists are patients. In the first act they entertain themselves by murdering their nurses, and in the second act they are revealed to be secret agents in the pay of rival intelligence services. I found the play amusing but at the same time irritating. These absurd creatures on the stage had no resemblance at all to any real physicist. I complained about the unreality of the characters to my friend Markus Fierz, a well-known Swiss physicist, who came with me to the play. "But don't you see?" said Fierz. "The whole point of the play is to show us how we look to the rest of the human race."

Fierz was right. The image of noble and virtuous dedication to truth, the image that scientists have traditionally presented to the public, is no longer credible. The public, having found out that the traditional image of the scientist as a secular saint is false, has gone to the opposite extreme and imagines us to be irresponsible devils playing with human lives. Dürrenmatt has held up the mirror to us and has shown us the image of ourselves as the public sees us. It is our task now to dispel these fantasies with facts, showing the public that scientists are neither saints nor devils but human beings sharing the common weaknesses of our species.

Historians who believe in the transcendence of science have portrayed scientists as living in a transcendent world of the intellect, superior to the transient, corruptible, mundane realities of the social world. Any scientist who claims to follow such exalted ideals is easily held up to ridicule as a pious fraud. We all know that scientists, like television evangelists and politicians, are not immune to the corrupting influences of power and money. Much of the history of science, like the history of religion, is a history of struggles driven by power and money. And yet this is not the whole story. Genuine saints occasionally play an important role, both in religion and in science. Einstein was an important figure in the history of science, and he was a firm believer in transcendence. For Einstein, science as a way of escape from mundane reality was no pretense. For many scientists less divinely gifted than Einstein, the chief reward for being a scientist is not the power and the money but the chance of catching a glimpse of the transcendent beauty of nature.

Both in science and in history there is room for a variety of styles and purposes. There is no necessary contradiction between the transcendence of science and the realities of social history. One may believe that in science nature will ultimately have the last word and still recognize an enormous role for human vainglory and viciousness in the practice of science before the last word is spoken. One may believe that the historian's job is to expose the hidden influences of power and money and still recognize that the laws of nature cannot be bent and cannot be corrupted by power and money. To my mind, the history of science is most illuminating when the frailties of human actors are put into juxtaposition with the transcendence of nature's laws.

Francis Crick is one of the great scientists of our century. He has recently published his personal narrative of the microbiological revolution that he helped to bring about, with a title borrowed from Keats, *What Mad Pursuit*. One of the most illuminating passages in his account compares two discoveries in which he was involved. One was the discovery of the double-helix structure of DNA, the other was the discovery of the triple-helix structure of the collagen molecule. Both molecules are biologically important, DNA being the carrier of genetic information, collagen being the protein that holds human bodies together. The two discoveries involved similar scientific techniques and aroused similar competitive passions in the scientists racing to be the first to find the structure.

Crick says that the two discoveries caused him equal excitement and equal pleasure at the time he was working on them. From the point of view of a historian who believes that science is a purely social construction, the two discoveries should have been equally significant. But in history as Crick experienced it, the two helixes were not equal. The double helix became the driving force of a new science, while the triple helix remained a footnote of interest only to specialists. Crick asks the question, how the different fates of the two helixes are to be explained. He answers the question by saying that human and social influences cannot explain the difference, that only the transcendent beauty of the double-helix structure and its genetic function can explain the difference. Nature herself, and not the scientist, decided what was important. In the history of the double helix, transcendence was real. Crick gives himself the credit for choosing an important problem to work on, but, he says, only Nature herself could tell how transcendentally important it would turn out to be.

My message is that science is a human activity, and the best way to understand it is to understand the individual human beings who practice it. Science is an art form and not a philosophical method. The great advances in science usually result from new tools rather than from new doctrines. If we try to squeeze science into a single philosophical viewpoint such as reductionism, we are like Procrustes chopping off the feet of his guests when they do not fit onto his bed. Science flourishes best when it uses freely all the tools at hand, unconstrained by preconceived notions of what science ought to be. Every time we introduce a new tool, it always leads to new and unexpected discoveries, because Nature's imagination is richer than ours.

Postscript, 2006

This essay was originally written as a lecture addressed to a meeting in 1992 that was supposed to discuss "the continuing primacy of reductionism as a key to understanding nature as we approach the twenty-first century." That explains why I devoted so much time to attacking reductionism. It turned out that many of the other participants at the meeting shared my views.

After the essay appeared in *The New York Review*, I received many good letters in response, some agreeing with me and some disagreeing. The best of them was from Saunders Mac Lane, a legendary figure in the world of mathematics. His letter and my reply were published in the October 5, 1995, issue of the *Review*. He objected vehemently to my statement that the later years of the great mathematician Hilbert were sterile. He had known Hilbert personally and professionally. His letter concludes, "Dyson simply does not understand reductionism and the deep purposes it can serve. Hilbert was not sterile." In my reply I said, "I too was exhilarated and inspired by the enormous deepening of mathematical understanding that grew in the 1930s out of the ruins of Hilbert's program of formalization. Only, Mac Lane would use the words 'upon the foundations' where I say 'out of the ruins.' Solid foundations and ruined hopes are not incompatible. Both were essential parts of the legacy that Hilbert left to his successors.... I do not deny the power and the beauty of reductionist science, as exemplified in the axioms and theorems of abstract algebra.... But I assert the equal power and beauty of constructive science, as exemplified in Gödel's construction of an undecidable proposition.... Hilbert himself was, of course, a master of both kinds of mathematics."

2

CAN SCIENCE BE ETHICAL?

ONE OF MY favorite monuments is a statue of Samuel Gompers not far from the Alamo in San Antonio, Texas. Under the statue is a quote from one of Gompers's speeches:

> What does labor want?
> We want more schoolhouses and less jails,
> More books and less guns,
> More learning and less vice,
> More leisure and less greed,
> More justice and less revenge,
> We want more opportunities to cultivate our better nature.

Samuel Gompers was the founder and first president of the American Federation of Labor. He established in America the tradition of practical bargaining between labor and management which led to an era of growth and prosperity for labor unions. Now, seventy years after Gompers's death, the unions have dwindled, while his dreams—more books and fewer guns, more leisure and less greed, more schoolhouses and fewer jails—have been tacitly abandoned. In a society without social justice and with a free-market ideology, guns, greed, and jails are bound to win.

When I was a student of mathematics in England fifty years ago, one of my teachers was the great mathematician G.H. Hardy, who wrote a little book, *A Mathematician's Apology*, explaining to the general public what mathematicians do. Hardy proudly proclaimed that his life had been devoted to the creation of totally useless works of abstract art, without any possible practical application. He had strong views about technology, which he summarized in the statement "A science is said to be useful if its development tends to accentuate the existing inequalities in the distribution of wealth, or more directly promotes the destruction of human life." He wrote these words while war was raging around him.

Still, the Hardy view of technology has some merit even in peacetime. Many of the technologies that are now racing ahead most rapidly, replacing human workers in factories and offices with machines, making stockholders richer and workers poorer, are indeed tending to accentuate the existing inequalities in the distribution of wealth. And the technologies of lethal force continue to be as profitable today as they were in Hardy's time. The marketplace judges technologies by their practical effectiveness, by whether they succeed or fail to do the job they are designed to do. But always, even for the most brilliantly successful technology, an ethical question lurks in the background: the question whether the job the technology is designed to do is actually worth doing.

The technologies that raise the fewest ethical problems are those that work on a human scale, brightening the lives of individual people. Lucky individuals in each generation find technology appropriate to their needs. For my father ninety years ago, technology was a motorcycle. He was an impoverished young musician growing up in England in the years before World War I, and the motorcycle came to him as a liberation. He was a working-class boy in a country dominated by the snobberies of class and accent. He learned to speak like a gentleman, but he did not belong in the world of gentlemen. The motorcycle was a great equalizer. On his motorcycle, he was the equal of a gentleman. He could make the grand tour of Europe without having inherited an upper-class income. He and three of his friends bought motorcycles and rode them all over Europe.

My father fell in love with his motorcycle and with the technical skills that it demanded. He understood, sixty years before Robert Pirsig wrote *Zen and the Art of Motorcycle Maintenance*, the spiritual quality of the motorcycle. In my father's day, roads were bad and repair shops few and far between. If you intended to travel any long distance, you needed to carry your own tool kit and spare parts and be prepared to take the machine apart and put it back together again. A breakdown of the machine in a remote place often required major surgery. It was as essential for a rider to understand the anatomy and physiology of the motorcycle as it was for a surgeon to understand the anatomy and physiology of a patient. It sometimes happened that my father and his friends would arrive at a village where no motorcycle had ever been seen before. When this happened, they would give rides to the village children and hope to be rewarded with a free supper at the village inn. Technology in the shape of a motorcycle was comradeship and freedom.

Fifty years after my father, I discovered joyful technology in the shape of a nuclear fission reactor. That was in 1956, in the first intoxicating days of peaceful nuclear energy, when the technology of reactors suddenly emerged from wartime secrecy and the public was invited to come and play with it. This was an invitation that I could not refuse. It looked then as if nuclear energy would be the great equalizer, providing cheap and abundant energy to rich and poor alike, just as fifty years earlier the motorcycle gave mobility to rich and poor alike in class-ridden England.

I joined the General Atomic Company in San Diego, where my friends were playing with the new technology. We invented and built a little reactor which we called the TRIGA, designed to be inherently safe. Inherent safety meant that it would not misbehave even if the people operating it were grossly incompetent. The company has been manufacturing and selling TRIGA reactors for forty years and is still selling them today, mostly to hospitals and medical centers, where they produce short-lived isotopes for diagnostic purposes. They have never misbehaved or caused any danger to the people who used them. They have only run into trouble in a few places where the neighbors objected to their presence on ideological grounds, no matter how safe they might be. We were successful with the TRIGA because it was designed to do a useful job at a price that a big hospital could afford. The price in 1956 was a quarter of a million dollars. Our work with the TRIGA was joyful because we finished it quickly, before the technology became entangled with politics and bureaucracy, before it became clear that nuclear energy was not and never could be the great equalizer.

Forty years after the invention of the TRIGA, my son George found another joyful and useful technology, the technology of CAD-CAM, computer-aided design and computer-aided manufacturing. CAD-CAM is the technology of the postnuclear generation, the technology that succeeded after nuclear energy failed. George is a boatbuilder. He designs seagoing kayaks. He uses modern materials to reconstruct the ancient craft of the Aleuts, who perfected their boats by trial and error over thousands of years and used them to travel prodigious distances across the northern Pacific. His boats are fast and rugged and seaworthy. When he began his boatbuilding twenty-five years ago, he was a nomad, traveling up and down the north Pacific coast, trying to live like an Aleut, and built his boats like an Aleut, shaping every part of each boat and stitching them together with his own hands. In those days he was a nature-child, in love with the wilderness, rejecting the urban society in which he had grown up. He built boats for his own use and for his friends, not as a commercial business.

As the years went by George made a graceful transition from the role of rebellious teenager to the role of solid citizen. He married, raised a daughter, bought a house in the city of Bellingham, and converted an abandoned tavern by the waterfront into a well-equipped workshop for his boats. His boats are now a business. And he discovered the joys of CAD-CAM.

His workshop now contains more computers and software than sewing needles and hand tools. It is a long time since he made the parts of a boat by hand. He now translates his designs directly into CAD-CAM software and transmits them electronically to a manufacturer who produces the parts. George collects the parts and sells them by mail order to his regular customers with instructions for assembling them into boats. Only on rare occasions, when a wealthy customer pays for a custom-built job, does George deliver a boat assembled in the workshop. The boat business occupies only a part of his time. He also runs a historical society concerned with the history and ethnography of the north Pacific. The technology of CAD-CAM has given George resources and leisure, so that he can visit the Aleuts in their native islands and reintroduce to the young islanders the forgotten skills of their ancestors.

Forty years into the future, which joyful new technology will be enriching the lives of our grandchildren? Perhaps they will be designing their own dogs and cats. Just as the technology of CAD-CAM began in the production lines of large manufacturing companies and later became accessible to individual citizens like George, the technology of genetic engineering may soon spread out from the biotechnology companies and agricultural industries and become accessible to our grandchildren. Designing dogs and cats in the privacy of a home may become as easy as designing boats in a waterfront workshop.

Instead of CAD-CAM we may have CAS-CAR, computer-aided selection and computer-aided reproduction. With the CAS-CAR software, you first program your pet's color scheme and behavior, and then transmit the program electronically to the artificial fertilization laboratory for implementation. Twelve weeks later, your pet is born, satisfaction guaranteed by the software company. When I recently described these possibilities in a public lecture at a children's museum in Vermont, I was verbally assaulted by a young woman in the audience. She accused me of violating the rights of animals. She said I was a typical scientist, one of those cruel people who spend their lives torturing animals for fun. I tried in vain to placate her by saying that I was only speaking of possibilities, that I was not actually myself engaged in designing dogs and cats. I had to admit that she had a legitimate complaint. Designing dogs and cats is an ethically dubious business. It is not as innocent as designing boats.

When the time comes, when the CAS-CAR software is available, when anybody with access to the software can order a dog with pink-and-purple spots that can crow like a rooster, some tough decisions will have to be made. Shall we allow private citizens to create dogs who will be objects of contempt and ridicule, unable to take their rightful place in dog society? And if not, where shall we draw the line between legitimate animal breeding and illegitimate creation of monsters? These are difficult questions that our children and grandchildren will have to answer. Perhaps I should have spoken to the audience in Vermont about designing roses and orchids instead of dogs and cats. Nobody seems to care so deeply for the dignity of roses and orchids. Vegetables, it seems, do not have rights. Dogs and cats are too close to being human. They have feelings like ours. If our grandchildren are allowed to design their own dogs and cats, the next step will be using the CAS-CAR software to design their own babies. Before that next step is reached, they ought to think carefully about the consequences.

What can we do today, in the world as we find it at the end of the twentieth century, to turn the evil consequences of technology into good? The ways in which science may work for good or evil in human society are many and various. As a general rule, to which there are many exceptions, science works for evil when its effect is to provide toys for the rich, and works for good when its effect is to provide necessities for the poor. Cheapness is an essential virtue. The motorcycle worked for good because it was cheap enough for a poor schoolteacher to own. Nuclear energy worked mostly for evil because it remained a toy for rich governments and rich companies to play with. "Toys for the rich" means not only toys in the literal sense but technological conveniences that are available to a minority of people and make it harder for those excluded to take part in the economic and cultural life of the community. "Necessities for the poor" include not only food and shelter but adequate public health services, adequate public transportation, and access to decent education and jobs.

The scientific advances of the nineteenth century and the first half of the twentieth were generally beneficial to society as a whole, spreading wealth to rich and poor alike with some degree of equity. The electric light, the telephone, the refrigerator, radio, television, synthetic fabrics, antibiotics, vitamins, and vaccines were social equalizers, making life easier and more comfortable for almost everybody, tending to narrow the gap between rich and poor rather than to widen it. Only in the second half of our century has the balance of advantage shifted. During the last forty years, the strongest efforts in pure science have been concentrated in highly esoteric fields remote from contact with everyday problems. Particle physics, low-temperature physics, and extragalactic astronomy are examples of pure sciences moving further and further away from their origins. The intensive pursuit of these sciences does not do much harm, or much good, to either the rich or the poor. The main social benefit provided by pure science in esoteric fields is to serve as a welfare program for scientists and engineers.

At the same time, the strongest efforts in applied science have been concentrated upon products that can be profitably sold. Since the rich can be expected to pay more than the poor for new products, market-driven applied science will usually result in the invention of toys for the rich. The laptop computer and the cellular telephone are the latest of the new toys. Now that a large fraction of high-paying jobs are advertised on the Internet, people excluded from the Internet are also excluded from access to jobs. The failure of science to produce benefits for the poor in recent decades is due to two factors working in combination: the pure scientists have become more detached from the mundane needs of humanity, and the applied scientists have become more attached to immediate profitability.

Although pure and applied science may appear to be moving in opposite directions, there is a single underlying cause that has affected them both. The cause is the power of committees in the administration and funding of science. In the case of pure science, the committees are composed of scientific experts performing the rituals of peer review. If a committee of scientific experts selects research projects by majority vote, projects in fashionable fields are supported while those in unfashionable fields are not. In recent decades, the fashionable fields have been moving further and further into specialized areas remote from contact with things that we can see and touch. In the case of applied science, the committees are composed of business executives and managers. Such people usually give support to products that affluent customers like themselves can buy.

Only a cantankerous man like Henry Ford, with dictatorial power over his business, would dare to create a mass market for automobiles by arbitrarily setting his prices low enough and his wages high enough that his workers could afford to buy his product. Both in pure science and in applied science, rule by committee discourages unfashionable and bold ventures. To bring about a real shift of priorities, scientists and entrepreneurs must assert their freedom to promote new technologies that are more friendly than the old to poor people and poor countries. The ethical standards of scientists must change as the scope of the good and evil caused by science has changed. In the long run, as Haldane and Einstein said, ethical progress is the only cure for the damage done by scientific progress.

The nuclear arms race is over, but the ethical problems raised by nonmilitary technology remain. The ethical problems arise from three "new ages" flooding over human society like tsunamis. First is the Information Age, already arrived and here to stay, driven by computers and digital memory. Second is the Biotechnology Age, due to arrive in full force early in the next century, driven by DNA sequencing and genetic engineering. Third is the Neurotechnology Age, likely to arrive later in the next century, driven by neural sensors and exposing the inner workings of human emotion and personality to manipulation. These three new technologies are profoundly disruptive. They offer liberation from ancient drudgery in factory, farm, and office. They offer healing of ancient diseases of body and mind. They offer wealth and power to the people who possess the skills to understand and control them. They destroy industries based on older technologies and make people trained in older skills useless. They are likely to bypass the poor and reward the rich. They will tend, as Hardy said eighty years ago, to accentuate the inequalities in the existing distribution of wealth, even if they do not, like nuclear technology, more directly promote the destruction of human life.

The poorer half of humanity needs cheap housing, cheap health care, and cheap education, accessible to everybody, with high quality and high aesthetic standards. The fundamental problem for human society in the next century is the mismatch between the three new waves of technology and the three basic needs of poor people. The gap between technology and needs is wide and growing wider. If technology continues along its present course, ignoring the needs of the poor and showering benefits upon the rich, the poor will sooner or later rebel against the tyranny of technology and turn to irrational and violent remedies. In the future, as in the past, the revolt of the poor is likely to impoverish rich and poor together.

The widening gap between technology and human needs can only be filled by ethics. We have seen in the last thirty years many examples of the power of ethics. The worldwide environmental movement, basing its power on ethical persuasion, has scored many victories over industrial wealth and technological arrogance. The most spectacular victory of the environmentalists was the downfall of the nuclear industry in the United States and many other countries, first in the domain of nuclear power and more recently in the domain of weapons. It was the environmental movement that closed down factories for making nuclear weapons in the United States, from plutonium-producing Hanford to warhead-producing Rocky Flats. Ethics can be a force more powerful than politics and economics.

Unfortunately, the environmental movement has so far concentrated its attention upon the evils that technology has done rather than upon the good that technology has failed to do. It is my hope that the attention of the Greens will shift in the next century from the negative to the positive. Ethical victories putting an end to technological follies are not enough. We need ethical victories of a different kind, engaging the power of technology positively in the pursuit of social justice.

If we can agree with Thomas Jefferson that these truths are self-evident, that all men are created equal, that they are endowed with certain inalienable rights, that among these are life, liberty, and the pursuit of happiness, then it should also be self-evident that the abandonment of millions of people in modern societies to unemployment and destitution is a worse defilement of the earth than nuclear power stations. If the ethical force of the environmental movement can defeat the manufacturers of nuclear power stations, the same force should also be able to foster the growth of technology that supplies the needs of impoverished humans at a price they can afford. This is the great task for technology in the coming century.

The free market will not by itself produce technology friendly to the poor. Only a technology positively guided by ethics can do it. The power of ethics must be exerted by the environmental movement and by concerned scientists, educators, and entrepreneurs working together. If we are wise, we shall also enlist in the common cause of social justice the enduring power of religion. Religion has in the past contributed mightily to many good causes, from the building of cathedrals and the education of children to the abolition of slavery. Religion will remain in the future a force equal in strength to science and equally committed to the long-range improvement of the human condition.

In the world of religion, over the centuries, there have been prophets of doom and prophets of hope, with hope in the end predominating. Science also gives warnings of doom and promises of hope, but the warnings and the promises of science cannot be separated. Every honest scientific prophet must mix the good news with the bad. Haldane was an honest prophet, showing us the evil done by science not as inescapable fate but as a challenge to be overcome. He wrote in his book *Daedalus* in 1923, "We are at present almost completely ignorant of biology, a fact which often escapes the notice of biologists, and renders them too presumptuous in their estimates of the present condition of their science, too modest in their claims for its future." Biology has made amazing progress since 1923, but Haldane's statement is still true.

We still know little about the biological processes that affect human beings most intimately—the development of speech and social skills in infants, the interplay between moods and emotions and learning and understanding in children and adults, the onset of aging and mental deterioration at the end of life. None of these processes will be understood within the next decade, but all of them might be understood within the next century. Understanding will then lead to new technologies that offer hope of preventing tragedies and ameliorating the human condition. Few people believe any longer in the romantic dream that human beings are perfectible. But most of us still believe that human beings are capable of improvement.

In public discussions of biotechnology today, the idea of improving the human race by artificial means is widely condemned. The idea is repugnant because it conjures up visions of Nazi doctors sterilizing Jews and killing defective children. There are many good reasons for condemning enforced sterilization and euthanasia. But the artificial improvement of human beings will come, one way or another, whether we like it or not, as soon as the progress of biological understanding makes it possible. When people are offered technical means to improve themselves and their children, no matter what they conceive improvement to mean, the offer will be accepted. Improvement may mean better health, longer life, a more cheerful disposition, a stronger heart, a smarter brain, the ability to earn more money as a rock star or baseball player or business executive. The technology of improvement may be hindered or delayed by regulation, but it cannot be permanently suppressed. Human improvement, like abortion today, will be officially disapproved, legally discouraged, or forbidden, but widely practiced. It will be seen by millions of citizens as a liberation from past constraints and injustices. Their freedom to choose cannot be permanently denied.

Two hundred years ago, William Blake engraved *The Gates of Paradise*, a little book of drawings and verses. One of the drawings, with the title *Aged Ignorance*, shows an old man wearing professorial eyeglasses and holding a large pair of scissors. In front of him, a winged child is running naked in the light from a rising sun. The old man sits with his back to the sun. With a self-satisfied smile he opens his scissors and clips the child's wings. With the picture goes a little poem:

> *In Time's Ocean falling drown'd,*
> *In Aged Ignorance profound,*
> *Holy and cold, I clip'd the Wings*
> *Of all Sublunary Things.*[1]

This picture is an image of the human condition in the era that is now beginning. The rising sun is biological science, throwing light of

1. *The Portable Blake*, edited by Alfred Kazin (Viking, 1946).

ever-increasing intensity onto the processes by which we live and feel and think. The winged child is human life, becoming for the first time aware of itself and its potentialities in the light of science. The old man is our existing human society, shaped by ages of past ignorance. Our laws, our loyalties, our fears and hatreds, our economic and social injustices, all grew slowly and are deeply rooted in the past. Inevitably the advance of biological knowledge will bring clashes between old institutions and new desires for human self-improvement. Old institutions will clip the wings of new desires. Up to a point, caution is justified and social constraints are necessary. The new technologies will be dangerous as well as liberating. But in the long run, social constraints must bend to new realities. Humanity cannot live forever with clipped wings. The vision of self-improvement, which William Blake and Samuel Gompers in their different ways proclaimed, will not vanish from the earth.

Postscript, 2006

Nine years later, the gap between rich and poor has grown wider. New technologies have continued to make stockholders richer and workers poorer. The main thesis of this essay, that technological progress does more harm than good unless it is accompanied by ethical progress, is even truer today than it was in 1997.

Only a few statements need to be corrected. The cell phone is no longer a toy for the rich but is becoming ubiquitous. I sat recently in the waiting room of the Social Security Administration office in Trenton, among a crowd of the poorer citizens of New Jersey, and was happy to see that many of them are now carrying cell phones. My son George continues to operate his boat business in Bellingham, but he is now better known as a writer and historian.

3

A MODERN HERETIC

THE FIRST TIME I met Thomas Gold was in 1946, when I served as a guinea pig in an experiment that he was doing on the capabilities of the human ear. Humans have a remarkable ability to discriminate the pitch of musical sounds. We can easily tell the difference when the frequency of a pure tone wobbles by as little as one percent. How do we do it? This was the question that Gold was determined to answer. There were two possible answers. Either the inner ear contains a set of finely tuned resonators that vibrate in response to incident sounds. Or the ear does not resonate but merely translates the incident sounds directly into neural signals that are then analyzed into pure tones by some unknown neural process inside our brains. In 1946 the professional physiologists who were experts in the anatomy and physiology of the ear believed that the second answer must be correct, that the discrimination of pitch happens in our brains and not in our ears. They rejected the first answer because they knew that the inner ear is a small cavity filled with flabby flesh and water. They could not imagine the flabby little membranes in the ear resonating like the strings of a harp or a piano.

Gold designed his experiment to prove the experts wrong. The experiment was simple, elegant, and original. During World War II he had been working for the Royal Navy on radio communications and radar. He built his apparatus out of war-surplus navy electronics and headphones. He fed into the headphones a signal consisting of short pulses of a pure tone, separated by intervals of silence. The silent intervals were at least ten times as long as the period of the pure tone. The pulses were all the same shape, but they had phases which could be reversed independently. To reverse the phase of a pulse means to reverse the movement of the speaker in the headphone. The speaker in a reversed pulse is pushing the air outward when the speaker in an unreversed pulse is pulling the air inward. Sometimes Gold gave all the pulses the same phase, and sometimes he alternated the phases so that the even pulses had one phase and the odd pulses had the opposite phase. All I had to do was to sit with the headphones on my ears and listen while Gold put in signals with either constant or alternating phases. I had to tell him from the sound whether the phase was constant or alternating.

When the silent interval between pulses was ten times the period of the pure tone, it was easy to tell the difference. I heard a noise like a mosquito, a hum and a buzz sounding together, and the quality of the hum changed noticeably when the phases were changed from constant to alternating. We repeated the trials with longer silent intervals. I could still detect the difference, when the silent interval was as long as thirty periods. I was not the only guinea pig. Several other friends of Gold listened to the signals and found similar results. The experiment showed that the human ear can remember the phase of a signal, after the signal stops, for thirty times the period of the signal. To be able to remember the phase, the ear must contain fine-tuned resonators that continue to vibrate during the intervals of silence. The result of the experiment proved that pitch discrimination is mainly done in the ear and not in the brain.

Besides having experimental proof that the ear can resonate, Gold also had a theory to explain how a fine-tuned resonator can be built out of flabby and dissipative materials. His theory was that the inner ear contains an electrical feedback system. The mechanical resonators are coupled to electrically powered sensors and drivers, so that the combined electromechanical system works like a finely tuned amplifier. The positive feedback provided by the electrical components counteracts the damping produced by the flabbiness of the mechanical components. Gold's experience as an electrical engineer made this theory seem plausible to him, although he could not identify the anatomical structures in the ear that functioned as sensors and drivers. In 1948 he published two papers, one reporting the results of the experiment and the other describing the theory.

Having myself participated in the experiment and listened to Gold explaining the theory, I never had any doubt that he was right. The professional auditory physiologists were equally sure that he was wrong. They found the theory implausible and the experiment unconvincing. They regarded Gold as an ignorant outsider intruding into a field where he had no training and no credentials. So for thirty years his work on hearing was ignored, and he moved on to other things.

Thirty years later, a new generation of auditory physiologists began to explore the ear with far more sophisticated tools. They discovered that everything that Gold had said in 1948 was true. The electrical sensors and drivers in the inner ear are now identified. They are two different kinds of hair cells, and they function in the way Gold said they should. The community of physiologists finally recognized the importance of his work, forty years after it was published.

Gold's study of the mechanism of hearing is typical of the way he has worked throughout his life. About once every five years, he invades a new field of research and proposes an outrageous theory that arouses intense opposition from the professional experts in the field. He then works very hard to prove the experts wrong. He does not always succeed. Sometimes it turns out that the experts are right and he is wrong. He is not afraid of being wrong. He was famously wrong at least twice, once when he promoted the theory of a steady-state universe in which matter is continuously created to keep the density constant as the universe expands, and once when he predicted that the moon would be covered with electrostatically supported dust into which the astronauts would sink as soon as they stepped onto the surface. When he is proved wrong, he concedes defeat with good humor. Science is no fun, he says, if you are never wrong. His wrong ideas are insignificant compared with his far more important right ideas. One of his important right ideas was the theory that pulsars, the regularly pulsing celestial radio sources discovered by radio astronomers in 1967, are rotating neutron stars. Unlike most of his right ideas, his theory of pulsars was accepted almost immediately by the experts.

Another of Gold's right ideas was rejected by the experts for an even longer time than his theory of hearing. This was his theory of the ninety-degree flip of the axis of rotation of the earth. In 1955 he published a revolutionary paper with the title "Instability of the Earth's Axis of Rotation." He proposed that the earth's axis might occasionally flip over through an angle of ninety degrees within a time of the order of a million years, so that the old north and south poles would move to the equator, and two points of the old equator would move to the poles. The flip would be triggered by movements of mass that would cause the old rotation axis to become unstable and the new rotation axis to become stable. For example, a large accumulation of ice at the old north and south poles might cause such an exchange of stability. Gold's paper was ignored by the experts for forty years. The experts at that time were focusing their attention narrowly on the phenomena of continental drift and the theory of plate tectonics. Gold's theory had nothing to do with continental drift or with plate tectonics, and it was therefore of no interest to them. The flip predicted by Gold would occur much more rapidly than continental drift, and would not change the positions of continents relative to one another. The flip would only change the positions of continents relative to the rotation axis.

In 1997 Joseph Kirschvink, an expert on rock magnetism at the California Institute of Technology, published a paper presenting evidence that a ninety-degree flip of the rotation axis actually occurred during a geologically short time in the Early Cambrian Era. This discovery is of great importance for the history of life, since the time of the flip appears to coincide with the time of the "Cambrian Explosion," the brief period when all the major varieties of higher organisms suddenly appear in the fossil record. It is possible that the flip of the rotation axis caused profound environmental changes in the oceans and triggered the rapid evolution of new life-forms. Kirschvink gives Gold credit for suggesting the theory that makes sense of his observations. If the theory had not been ignored for forty years, the evidence that confirms it might have been collected sooner.

Gold's most controversial idea is the nonbiological origin of natural gas and oil. He advocates a theory that natural gas and oil come from reservoirs deep in the earth and are relics of the material out of which the earth condensed. The biological molecules found in oil show that the oil is contaminated by living creatures, not that the oil was produced by living creatures. This theory, like his theories of hearing and of polar flip, contradicts the entrenched dogma of the experts. Once again, Gold is regarded as an intruder ignorant of the field that he is invading. In fact, Gold is an intruder but he is not ignorant. He knows the details of the geology and chemistry of natural gas and oil. His arguments supporting his theory are based on a wealth of factual information. Perhaps it will once again take us forty years to decide whether the theory is right. Whether the theory of nonbiological origin is ultimately found to be right or wrong, the collection of evidence to test it will add greatly to our knowledge of the earth and its history.

Finally, the most recent of Gold's revolutionary proposals is the subject of his book *The Deep Hot Biosphere*.[1] His theory says that

1. Springer-Verlag, 1999.

the entire crust of the earth, down to a depth of several miles, is populated with living creatures. The creatures that we see living on the surface are only a small part of the biosphere. The greater and more ancient part of the biosphere is deep and hot. The theory is supported by a considerable mass of evidence. I do not need to summarize the evidence here, because it is clearly presented in the book. I prefer to let Gold speak for himself. The purpose of my foreword is only to explain how the theory of the deep hot biosphere fits into the general pattern of Gold's life and work. Gold's theories are always original, always important, usually controversial, and usually right. It is my belief, based on fifty years of observation of Gold as a friend and colleague, that the deep hot biosphere is all of the above: original, important, controversial, and right.

Postscript, 2006

Thomas Gold died in June 2004. Shortly before he died, an experiment was done at the Carnegie Institution of Washington Geophysical Laboratory to test his theory that natural gas is generated deep in the earth's mantle.[2] The experiment, carried out with tiny quantities of mantle materials exposed to high temperature and pressure in a diamond anvil cell, demonstrated abundant production of methane. The authors sent a message to Gold to tell him that his theory had been confirmed, only to learn that he had died three days earlier.

2. H.P. Scott et al., "Generation of Methane in the Earth's Mantle: *In Situ* High Pressure–Temperature Measurements of Carbonate Reduction," *Proceedings of the National Academy of Sciences*, Vol. 101, No. 39 (September 28, 2004), pp. 14023–14026.

4

THE FUTURE NEEDS US

PREY[1] IS A thriller, well constructed and fun to read, like Michael Crichton's other books. The main characters are the narrator, Jack, and his wife, Julia, parents of three lively children, successfully combining the joys of parenthood with the pursuit of brilliant careers in the high-tech world of Silicon Valley. Julia works for a company called Xymos that is developing nanorobots, tiny machines that can move around and function autonomously but are programmed to work together like an army of ants. Jack works for a company called MediaTronics that makes software to coordinate the actions of large groups of autonomous agents. His programs give intelligence and flexibility to her machines.

Things start to go wrong when Jack loses his job and is left to take care of the kids, while Julia is working longer and longer hours at her laboratory and losing interest in the family. She is engaged in a secret struggle to develop her nanorobots into a stealthy photo-reconnaissance system that can be sold to the United States Army. To increase the power and performance of the system, she incorporates living bacteria into the nanorobots so that they can reproduce and evolve rapidly. She reprograms them with Jack's newest autonomous-agent software so that they can learn from experience.

1. HarperCollins, 2002.

Even with these improvements the nanorobots fail to meet the army's specifications, and Xymos loses its army funding. After that, Julia desperately tries to convert the photo-reconnaissance system into a medical diagnostic system that can be sold on the civilian market. Her idea is to train the nanorobots to enter and explore the human body, so that they can locate tumors and other pathological conditions more precisely than can be done with X-rays and ultrasound working from the outside.

Experimenting with the medical applications of her nanorobots, she uses herself as a guinea pig and becomes chronically infected. The nanorobots learn how to establish themselves as symbionts within her body, and then gradually gain control over her mind. In her deranged state, she deliberately infects three of her colleagues at the laboratory with nanorobots. She also lets a swarm of nanorobots loose into the environment where they prey upon wildlife and rapidly increase in numbers.

The main part of the story concerns Jack's slow realization that something is seriously amiss with his wife and with the project in which she is engaged. Only at the end does he understand the full horror of her transformation. With the help of a loyal young woman friend, he confronts Julia and douses her with a spray of bacteriophage that is lethal to the bacteria inside her. But Julia and her infected colleagues are no longer able to survive without the symbiotic nanorobots that have taken over their minds. Under the spray of bacteriophage they collapse and die, like the Wicked Witch of the West in *The Wizard of Oz* when Dorothy throws a bucket of water over her. After Julia's demise, Jack and his girlfriend finish the job of destroying the nanorobots inside and outside the laboratory with fire and high explosives. In the final scene, Jack is back with his kids, wondering whether the nanorobots are gone for good, or whether the Xymos corporation may still be developing other nanorobot projects that will turn into nightmares.

What are we to make of this fairy story? There are two ways to look at it. On the one hand, we may enjoy it as a story and not worry whether some parts of it might come true. On the other hand, we may read it as an urgent warning of dangers lying ahead if present-day technological developments are allowed to continue. The author says plainly, in an introductory chapter with the title "Artificial Evolution in the Twenty-first Century," that he intends his story to be taken seriously.

It is easy to demonstrate that the details of the story are technically flawed. Consider for example the size of the nanorobots. In a commercial presentation advertising the Xymos medical diagnostic system, Julia says, "We can do all this because the camera is smaller than a red blood cell." The camera is one of her nanorobots. It must be as small as that, since Julia describes it swimming in the human bloodstream inside the capillaries that carry blood through the lungs. The capillaries are only just wide enough for red blood cells to pass through. But later in the book Jack encounters swarms of nanorobots chasing him in the open air like a swarm of ants or bees. These nanorobots are flying through the air as fast as he can run. Fortunately for Jack and unfortunately for the story, the laws of physics do not allow very small creatures to fly fast. The viscous drag of air or water becomes stronger as the creature becomes smaller. Flying through air, for a nanorobot the size of a red blood cell, would be like swimming through molasses for a human being. Roughly speaking, the top speed of a swimmer or flyer is proportional to its length. A generous upper limit to the speed of a nanorobot flying through air or swimming through water would be a tenth of an inch per second, barely fast enough to chase a snail. For nanorobots to behave like a swarm of insects, they would have to be as large as insects.

Other technical flaws in the story are easy to find. The swarms of nanorobots flying in the open air are said to be powered by solar energy. But the solar energy falling onto their very small area is insufficient to power their movements, even if we credit them with a magical ability to use solar energy with 100 percent efficiency. I could continue with a list of technical details that are scientifically impossible for one reason or another, but that would miss the main point of the story. The story is about human beings and not about nanorobots. The main point is that Julia is a credible human being. She is a capable and well-meaning woman in a responsible position, with the fate of a company resting on her shoulders. She decides that the only way to save the company from bankruptcy is to push ahead with a risky technology. Unable to face the failure of her company and her career, she continues with her experiments regardless of the risks. She is a gambler playing for such high stakes that she cannot afford to lose. In the end she loses not only her company and her career but her family and her life. It is a credible human story, and in the end the technical details do not matter.

This story reminds me of Nevil Shute's *On the Beach*, published in 1957, a novel describing the extinction of mankind by radiological warfare. Shute's poignant translation of apocalyptic disaster into the everyday voices of real people caught the imagination of the world. His book became an international best seller and was made into a successful film. The book and the film created an enduring myth, a myth which entered consciously or subconsciously into all subsequent thinking about nuclear war. The myth pictures nuclear war as silent inexorable death from which there is no escape, with radioactive cobalt sweeping slowly down the sky from the northern to the southern hemisphere. The people of Australia, after the northern hemisphere is dead, live out their lives quietly and bravely to the end. The Australian government provides a supply of euthanasia pills for citizens to use when the symptoms of radiation sickness become unpleasant. Parents are advised to give the pills to their children first before they become sick. There is no hope of survival; there is no talk of building an underground Noah's Ark to keep earth's creatures alive until the cobalt decays. Shute imagined the human species calmly acquiescing in its extinction.

The myth of *On the Beach* is technically flawed in many ways. Almost all the details are wrong: radioactive cobalt would not substantially increase the lethality of large hydrogen bombs; fallout would not descend uniformly over large areas but would fall sporadically in space and time; people could protect themselves from the radioactivity by sheltering under a few feet of dirt; and the war is supposed to have happened in 1961, too soon for even the most malevolent country to have acquired the megatonnage needed to give a lethal dose of radiation to the entire earth. Nevertheless, the myth did what Shute intended it to do. On the fundamental human level, in spite of the technical inaccuracies, it spoke truth. It told the world, in language that everyone could understand, that nuclear war means death. And the world listened.

Prey is not as good as *On the Beach*, but it is bringing us an equally important message. The message is that biotechnology in the twenty-first century is as dangerous as nuclear technology in the twentieth. The dangers do not lie in any particular gadgets such as nanorobots or autonomous agents. The dangers arise from knowledge, from our inexorably growing understanding of the basic processes of life. The message is that biological knowledge irresponsibly applied means death. And we may hope that the world will listen.

From this point on, I assume that the basic message of *Prey* is true. I assume that the growth of biological knowledge during the century now beginning will bring grave dangers to human society and to the ecology of our planet. The rest of this review is concerned with the question of what we should do to mitigate the dangers. What is the appropriate response to dangers that are hypothetical and poorly understood? In this matter, as in other situations where public health hazards and environmental risks must be assessed and regulated, there are two strongly opposed points of view. One point of view is based on the "precautionary principle." The precautionary principle says that when there is any risk of a major disaster, no action should be permitted that increases the risk. If, as often happens, an action promises to bring substantial benefits together with some risk of a major disaster, no balancing of benefits against risks is to be allowed. Any action carrying a risk of major disaster must be prohibited, regardless of the costs of prohibition.

The opposing point of view holds that risks are unavoidable, that no possible course of action or inaction will eliminate risks, and that a prudent course of action must be based on a balancing of risks against benefits and costs. In particular, when any prohibition of dangerous science and technology is contemplated, one of the costs that must be considered is the cost to human freedom. I call the first point of view precautionary and the second point of view libertarian. In April 2000, Bill Joy, co-founder and chief scientist at Sun Microsystems, a large and successful computer company, published an article in *Wired* magazine with the title "Why the Future Doesn't Need Us," and the subtitle "Our most powerful 21st-century technologies—robotics, genetic engineering, and nanotech—are threatening to make humans an endangered species." It was a big surprise to see one of the leaders of high-tech industry arguing passionately for a slowing down of technology that might become dangerous. Bill Joy became a spokesman for the precautionary view.

Nine months later, in January 2001, the annual meeting of the World Economic Forum was held in Davos, Switzerland. Most of the people at the forum are captains of industry, presidents of foundations, or government officials. But in 2001 they decided to invite some scientists and writers and artists to add some intellectual sparkle to the meeting. Bill and I were both invited and asked to debate the question: Is our technology out of control? Bill was taking an extreme position on the precautionary side, and I was asked to take an extreme position on the libertarian side, to make the debate interesting.

In what follows I shall summarize our debate.[2] To be sure that I am not misrepresenting Bill, I quote here only from his published writings.

My first quote is from Bill's article in *Wired*:

> The 21st-century technologies—genetics, nanotechnology, and robotics (GNR)—are so powerful that they can spawn whole new classes of accidents and abuses. Most dangerously, for the first time, these accidents and abuses are widely within the reach of individuals or small groups. They will not require large facilities or rare raw materials. Knowledge alone will enable the use of them.
>
> Thus we have the possibility not just of weapons of mass destruction but of knowledge-enabled mass destruction (KMD), this destructiveness hugely amplified by the power of self-replication.
>
> I think it is no exaggeration to say we are on the cusp of the further perfection of extreme evil, an evil whose possibility spreads well beyond that which weapons of mass destruction bequeathed to the nation-states, on to a surprising and terrible empowerment of extreme individuals.

This was written a year and a half before the events of September 2001. I don't know whether Bill at that time had Osama bin Laden in mind. He certainly had in mind the possibility of a Unabomber taking his revenge on society with genetically engineered microbes rather than with chemical explosives.

Second quote. Here Bill is quoting Eric Drexler, the chief prophet

2. This description of my debate with Bill Joy is taken from a lecture that I gave at the University of Virginia in 2004. The lecture will be published in a forthcoming book, *A Many-Colored Glass: Reflections on the Place of Life in the Universe* (University of Virginia Press, 2006).

of nanotechnology. Drexler set up the Foresight Institute to promote the benign uses of nanotechnology and to warn against the dangerous uses. Here is Drexler:

> Tough omnivorous [synthetic] "bacteria" could out-compete real bacteria: They could spread like blowing pollen, replicate swiftly, and reduce the biosphere to dust in a matter of days. Dangerous replicators could easily be too tough, small, and rapidly spreading to stop—at least if we make no preparation. We have trouble enough controlling viruses and fruit flies....
>
> We cannot afford certain kinds of accidents with replicating assemblers.

The idea of nanotechnology is to build machines on a tiny scale that are as capable as living cells, but made of different materials so that they are more rugged and more versatile. One kind of nanomachine is the assembler, which is a tiny factory that can manufacture other machines, including replicas of itself. Drexler understood from the beginning that a replicating assembler would be a tool of immense power for good or for evil. Fortunately or unfortunately, nanotechnology has moved more slowly than Drexler expected. Nothing remotely resembling an assembler has yet emerged. The most useful products of nanotechnology so far are computer chips. They have no capacity for replicating either themselves or anything else.

My last quote from Bill comes from an article he published in *The Washington Post*, summing up the dangers that he foresees and recommending a program of action to avoid them:

> We who are involved in advancing the new technologies must devote our best efforts to heading off disaster. I offer here a list of first steps suggested by our history with weapons of mass destruction:

(1) Have scientists and technologists (and corporate leaders as well) take a vow, along the lines of the Hippocratic Oath, to avoid work on potential and actual weapons of mass destruction....

(2) Create an international body to publicly examine the dangers and ethical issues of new technology....

(3) Use stricter notions of liability, forcing companies to take responsibility for consequences through a private-sector mechanism—insurance....

(4) Internationalize control of knowledge and technologies that have great potential but are judged too dangerous to be made commercially available....

(5) Relinquish pursuit of that knowledge and development of those technologies so dangerous that we judge it better that they never be available. I too believe in the pursuit of knowledge and development of technologies; yet, we already have seen cases, such as biological weapons, where relinquishment is the obvious wise choice.

Next comes my response to Bill. I agreed that the dangers he described are real, but I disagreed with some details of his argument, and I disagreed strongly with his remedies. I began by speaking about the history of biological weapons and gene-splicing experiments, and the successes and failures of efforts to regulate them. Bill ignores the long history of effective action by the international biological community to regulate and prohibit dangerous technologies. Gene-splicing experiments began in many countries when the technique of sticking pieces of DNA together was discovered in 1975. Two leading biologists, Maxine Singer and Paul Berg, issued a call for a moratorium on all such experiments until the dangers could be carefully assessed. There were obvious dangers to public health, for example if genes for deadly toxins could be inserted into bacteria that are normally endemic in human populations. Biologists all over the world quickly agreed to the moratorium, and experiments were halted everywhere for ten months. During the ten months, two international conferences were held to work out the guidelines for permissible and forbidden experiments. The guidelines established rules of physical and biological containment for permitted experiments involving various degrees of risk. The most dangerous experiments were forbidden outright. These guidelines were adopted voluntarily by the biologists and have been observed ever since, with changes made from time to time in response to new discoveries. As a result, no serious health hazards have arisen from the experiments in twenty-five years. This is a shining example of responsible citizenship, showing that it is possible for scientists to protect the public from injury while preserving the freedom of science.

The history of biological weapons is a more complicated story. The United States, Britain, and the Soviet Union all had large programs to develop and stockpile biological weapons during and after the Second World War. But these were low-key efforts compared with the programs to develop nuclear weapons. Unlike the well-known physicists who pushed the nuclear bomb programs ahead with great enthusiasm, the biologists never pushed hard for biological weapons. The great majority of biologists had nothing to do with weapons. The few biologists who were involved with the weapons program were mostly opposed to it.

The strongest of the opponents in the United States was Matthew Meselson, who had the good luck to be a neighbor and friend of Henry Kissinger in 1968 when Nixon became president. Kissinger became national security adviser to President Nixon. Meselson seized the opportunity to convince Kissinger, and Kissinger convinced Nixon, that the American biological weapons program was far more dangerous to the United States than to any possible enemy. On the one hand, it was difficult to imagine any circumstances in which the

United States would wish to use these weapons, and on the other hand, it was easy to imagine circumstances in which some of the weapons could fall into the hands of terrorists.

So Nixon in 1969 boldly declared that the United States was dismantling the entire program and destroying the stockpile of weapons. This was a unilateral move, not requiring any international agreement or ratification by the American Senate. The development of weapons was duly stopped and the weapons were destroyed. Britain quickly followed suit. In 1972, as a result of Nixon's initiative, an international convention was signed by the US, the UK, and the USSR, imposing a permanent prohibition of biological weapons on all three countries. Many other countries subsequently signed the convention.

As we now know, the Soviet Union violated the Biological Weapons Convention of 1972 on an extensive scale, continuing to develop new weapons and to accumulate stockpiles until its collapse in 1991. After the collapse, Russia declared its adherence to the convention and announced that the Soviet program had now finally been stopped. But many of the old Soviet research and production centers remain hidden behind walls of secrecy, and Russia has never provided the world with convincing evidence that the program is not continuing. It is quite possible that stockpiles of biological weapons continue to exist in Russia and in other countries. Nevertheless, the 1972 convention remains legally in force and most countries have signed it. Even if the convention is unverifiable and even if it is violated, we are far better off with it than without it. Without the convention, we would not have any legal ground for complaint or for preventive action whenever a biological weapons program anywhere in the world is discovered. With the convention, the danger of biological weapons is not eliminated but it is significantly reduced. Again, biologists in general and Meselson in particular deserve credit for making this happen in the real world of national politics and international rivalries.

The last part of my reply to Bill Joy concerns remedies for the dangers that we all agree exist. Bill says, "Internationalize control of knowledge" and "Relinquish pursuit of that knowledge . . . so dangerous that we judge it better that [it] never be available." Bill is advocating censorship of scientific inquiry, either by international or national authorities. I am opposed to this kind of censorship. It is often said that the risks of modern biotechnology are historically unparalleled because the consequences of letting a new living creature loose in the world may be irreversible. I think we can find a good historical parallel where a government was trying to guard against dangers that were equally irreversible.

Three hundred and fifty-nine years ago, the poet John Milton wrote a speech with the title *Areopagitica*, addressed to the Parliament of England. He was arguing for the liberty of unlicensed printing. I am suggesting that there is an analogy between the seventeenth-century fear of moral contagion by soul-corrupting books and the twenty-first-century fear of physical contagion by pathogenic microbes. In both cases, the fear was neither groundless nor unreasonable. In 1644, when Milton was writing, England was engaged in a long and bloody civil war, and the Thirty Years' War, which devastated Germany, had four years still to run. These seventeenth-century wars were religious wars, in which differences of doctrine played a great part. In that century, books not only corrupted souls but also mangled bodies. The risks of letting books go free into the world were rightly regarded by the English Parliament as potentially lethal as well as irreversible. Milton argued that the risks must nevertheless be accepted. I believe his message may still have value for our own times, if the word "books" is replaced by the word "experiments." Here is Milton:

> I deny not, but that it is of greatest concernment in the Church and Commonwealth, to have a vigilant eye how books demean themselves as well as men; and thereafter to confine, imprison, and do sharpest justice on them as malefactors. . . . I know they

> are as lively, and as vigorously productive, as those fabulous dragon's teeth; and being sown up and down, may chance to spring up armed men.

The important word in Milton's statement is "thereafter." Books should not be convicted and imprisoned until after they have done some damage. What Milton declared unacceptable was prior censorship, prohibiting books from ever seeing the light of day. Next, Milton comes to the heart of the matter, the difficulty of regulating "things, uncertainly and yet equally working to good and to evil":

> Suppose we could expel sin by this means; look how much we thus expel of sin, so much we expel of virtue: for the matter of them both is the same; remove that, and ye remove them both alike.
>
> This justifies the high providence of God, who, though he commands us temperance, justice, continence, yet pours out before us even to a profuseness all desirable things, and gives us minds that can wander beyond all limit and satiety. Why should we then affect a rigor contrary to the manner of God and of nature, by abridging or scanting those means, which books freely permitted are, both to the trial of virtue, and the exercise of truth. It would be better done to learn that the law must needs be frivolous which goes to restrain things, uncertainly and yet equally working to good and to evil.

My last quotation expresses Milton's patriotic pride in the intellectual vitality of seventeenth-century England, a pride that twenty-first-century Americans have good reason to share:

> Lords and Commons of England, consider what Nation it is whereof ye are, and whereof ye are the governors: a Nation not

> slow and dull, but of a quick, ingenious, and piercing spirit, acute to invent, subtle and sinewy to discourse, not beneath the reach of any point the highest that human capacity can soar to.... Nor is it for nothing that the grave and frugal Transylvanian sends out yearly from as far as the mountainous borders of Russia, and beyond the Hercynian wilderness, not their youth, but their staid men, to learn our language and our theologic arts.

Perhaps, after all, as we struggle to deal with the enduring problems of reconciling individual freedom with public safety, the wisdom of a great poet who died more than three hundred years ago may still be helpful.

That was the end of the debate. No vote was taken to determine who won. The purpose of the debate was not to win but to educate. Bill Joy and I remain friends.

5

WHAT A WORLD!

IT IS REFRESHING to read a book full of facts about our planet and the life that has transformed it, written by an author who does not allow facts to be obscured or overshadowed by politics. Vaclav Smil is well aware of the political disputes that are now raging about the effects of human activities on climate and biodiversity, but in *The Earth's Biosphere: Evolution, Dynamics, and Change*[1] he does not give them more attention than they deserve. He emphasizes the enormous gaps in our knowledge, the sparseness of our observations, and the superficiality of our theories. He calls attention to the many aspects of planetary evolution that are poorly understood, and that must be better understood before we can reach an accurate diagnosis of the present condition of our planet. When we are trying to take care of a planet, just as when we are taking care of a human patient, diseases must be diagnosed before they can be cured.

The book has two themes, a major and a minor one. The major theme is the description of the biosphere. The biosphere is the interacting web of plants and rocks, fungi and soils, animals and oceans, microbes and air that constitutes the habitat of life on our planet. To understand the biosphere, it is essential to see it from both sides, from

1. MIT Press, 2002.

below as a multitude of details and from above as a single integrated system. This book gives a comprehensive account of biological details and a summary of the global cycles of matter and energy that tie the system together. Every detail and every cycle is documented with references to the technical literature. There are forty pages of bibliography, containing more than a thousand references, ranging from John Ray's 1686 *History of Plants* to the 2001 report of the Intergovernmental Programme on Climatic Change. The bibliography will make this book a useful work of reference for students and teachers. The text is also intended to be read by ordinary citizens who are not students or teachers but have a serious interest in environmental problems.

The minor theme of the book is the life and work of Vladimir Vernadsky. Vernadsky did not invent the word "biosphere," but he was the first to make it a central concept unifying the study of the earth with the study of life. In Russia he is honored as one of the leading figures of twentieth-century science, while in the West his name is hardly known. Vaclav Smil, who is himself a bridge between East and West, educated in Prague and living in Canada, uses this book as an opportunity to bring Vernadsky to life and to make the West aware of his ideas. Every chapter begins with a quotation from Vernadsky's book *The Biosphere*, which summarized his thinking and was written for a wide audience. The first chapter, with the title "Evolution of the Idea," begins with Vernadsky saying, "A new character is imparted to the planet by this powerful cosmic force. The radiations that pour upon the Earth cause the biosphere to take on properties unknown to lifeless planetary surfaces, and thus transform the face of the Earth." The last chapter, with the title "Civilization and the Biosphere," begins with the quotation "Man, alone, violates the established order."

The meaning of this last quotation becomes clearer when we place it in its context. Man violates the established order not only by burning coal and oil but by farming and weeding. This is what Vernadsky wrote:

> In cultivated areas it is only at the expense of great effort that civilized man can secure crops unmixed with weeds which spring up everywhere. Before man appeared on the Earth, the vegetation everywhere must have reached its maximum possible development, a state of equilibrium, attained through centuries of growth. Such a state can be seen in the virgin steppes which still exist in parts of Russia.... As far as the eye could see, there was nothing but the waist-high growth of feather-grass, a continuous clothing to the Earth, protecting it against the heat of the sun. Moss and lichen, profiting by the conservation of the moisture in the soil, remained green throughout the heat of summer under the shadow of the leaves.
>
> Man, alone, violates the established order and, by cultivation, upsets the equilibrium.... He sees this when he is obliged to oppose the pressure of life in defending against this invader fields which he wants to cultivate. He sees it, too, if he watches the surrounding world of nature with attentive eyes; the secret, silent, inexorable fight for existence waged all around him by green vegetation. Sensing this movement, he may experience the reality of the assault of the forest on the steppe-land, or the gradual suffocation of the forest by the rising tide of lichens from the tundra.

In these words we hear the authentic voice of Vernadsky, talking like the doctor Mikhail Astrov in Chekhov's play *Uncle Vanya*. His statement of the facts is scientifically accurate, but is expressed in the language of drama and poetry. Vernadsky and Chekhov were contemporaries. Both belonged to the circle of philosophizing intellectuals that Chekhov portrays so poignantly in his plays. Vernadsky was a Chekhov character who also happened to be a world-class scientist.

Vernadsky was a geochemist, born in 1863 in Kiev, the son of a professor of political economy. In 1889 he worked as a student with

Pierre Curie in Paris, and in 1902 he became a full professor at Moscow University. After the first Russian revolution of 1905, which forced the Tsar to give some share in the affairs of government to a representative assembly called the Duma, Vernadsky was an important political figure. He was one of the founders of the Constitutional Democratic Party, generally known by its acronym Kadet. The Kadet party tried to provide the loyal opposition that Russia desperately needed in order to achieve far-reaching political reform without bloodshed. Unfortunately, the majority of intellectuals supported social revolutionary parties and did not believe in gradual reform.

Through the years from 1908 to 1918, Vernadsky remained a member of the central committee of the Kadet party, struggling to establish democratic government in Russia against bitter opposition from the Tsar's bureaucrats on the right and the social revolutionaries on the left. After the Bolshevik Revolution, most of the Kadet leaders were executed. Vernadsky was spared because he was a famous scientist and had some friends in Lenin's inner circle, but his political life was over. He spent some years as an exile in Paris, giving lectures at the Sorbonne on geochemistry and writing his book *The Biosphere*. In 1926, at the age of sixty-two, he returned peacefully to Russia and published the book in Leningrad. He refused to join the Communist Party, but continued to live until his death in 1945 as one of the respected elder statesmen of Soviet science.

In Russia the disciplines of geochemistry and biology remained unified, with Vernadsky's vision of the biosphere as a central theme. After Vernadsky's death, his books and papers continued to be read and studied. Russian biologists aimed to understand life by integrating it into ecological communities and planetary processes. Meanwhile, in the West, biology developed in a strongly reductionist direction, the aim being to understand life by reducing it to genes and molecules. Reductionist biology was enormously successful and came to dominate the thinking of Western biologists.

There is in fact no incompatibility between reductionist and integrative biology. Genes and molecules and ecologies and biospheres are all essential parts of the world we live in. To understand our world fully, both kinds of biology are needed. If science had been uncontaminated by politics, the reductionist and integrative approaches to biology in the West and the East would have blended together during Vernadsky's lifetime and merged into a balanced view of the biosphere. But in the 1930s, biology in the Soviet Union was almost destroyed by Trofim Lysenko's murderous campaign against Mendelian genetics. In Russia reductionist biology was forbidden, and in the West the Russian tradition of integrative biology was discredited because Lysenko appeared to approve of it. In the West Vernadsky's ideas were ignored and his books were unread. A complete translation of *The Biosphere* into English was only published in 1998.[2] After seventy years of dominance of reductionist biology, Vernadsky's language now seems quaint and old-fashioned.

One of the great might-have-beens of history is the world that would have emerged if the statesmen of Europe had had the wisdom to deal peacefully with the Serbian crisis of 1914. If World War I had never happened, the rapid economic growth that Russia experienced from 1905 to 1914 would probably have continued. The Bolsheviks would probably have remained a small group of outlaws without any wide following, and would not have had an opportunity to seize power. The Tsar's government might have evolved into a constitutional monarchy, and the Kadet party might have emerged as the leader of a liberal parliamentary regime. In that imaginary world, Vernadsky might have been prime minister of Russia, guiding his country along the path of economic and scientific development, ending with full integration into the world community. After reading some of his writings, I have little doubt that he would have chosen to stay in politics if he had had the chance. He would not then have had time to

2. V. I. Vernadsky, *The Biosphere*, translated by D. B. Langmuir (Copernicus, 1998).

resume his work as a scientist and write *The Biosphere.* Instead of being the founder of a new discipline of science, he might have been the savior of his country.

From Vernadsky and his dreams, I turn now to the major theme of Smil's book, which is the difficulty of understanding the behavior of the biosphere on a global scale. Even the nonliving processes governing weather and climate are difficult to understand. The living processes governing the fertility of forests and oceans are even more difficult. As an example to illustrate the difficulties, I look at the effects on the biosphere of carbon dioxide in the atmosphere. This is one of the subjects in Smil's book, but it is the reviewer, not the author, who is responsible for giving it emphasis here. As a result of the burning of coal and oil, the driving of cars, and other human activities, the carbon dioxide in the atmosphere is increasing at a rate of about half a percent per year.

Everyone agrees that the increasing abundance of carbon dioxide has two important consequences. First, carbon dioxide is a greenhouse gas, transparent to sunlight but partially opaque to the heat radiation that transports energy from the earth's surface into space. Second, carbon dioxide is an essential nutrient for plants on land and in the ocean. The increase in carbon dioxide causes changes, both in the transport of energy through the atmosphere and in the growth and reproduction of plants. Opinions differ on two crucial questions. Are the physical or the biological effects of carbon dioxide more important? Are the effects, either separately or together, beneficial or harmful? In his last two chapters, Smil summarizes the evidence bearing on these questions, but does not presume to answer them.

The physical effects of carbon dioxide are seen in changes of rainfall, cloudiness, wind strength, and temperature, which are customarily lumped together in the misleading phrase "global warming." This phrase is misleading because the warming caused by the greenhouse effect of increased carbon dioxide is not evenly distributed. In humid air, the effect of carbon dioxide on the transport of heat by radiation is less important, because it is outweighed by the much larger greenhouse effect of water vapor. The effect of carbon dioxide is more important where the air is dry, and air is usually dry only where it is cold. The warming mainly occurs where air is cold and dry, mainly in the arctic rather than in the tropics, mainly in winter rather than in summer, and mainly at night rather than in daytime. The warming is real, but it is mostly making cold places warmer rather than making hot places hotter. To represent this local warming by a global average is misleading, because the global average is only a fraction of a degree while the local warming at high latitudes is much larger. Also, local changes in rainfall, whether they are increases or decreases, are usually more important than changes in temperature. It is better to use the phrase "climate change" rather than "global warming" to describe the physical effects of carbon dioxide.

The biological effects of carbon dioxide on plants can be seen in changes of rate of growth, ratio of roots to shoots, and water requirements, which are different for different species and may result in shifts of the ecological balance from one kind of plant community to another. Effects on plant communities will also cause effects on dependent communities of microbes and animals. Biological effects are difficult to measure but are likely to be large. Experiments in greenhouses with an atmosphere enriched in carbon dioxide show that the yields of many crop plants increase roughly with the square root of the carbon dioxide abundance. If this were true for the major crop plants grown in the open air, it would mean that the 30 percent increase in carbon dioxide produced by fossil fuel–burning over the last sixty years would have resulted in a 15 percent increase of the world's food supply. A similar increase might have occurred in the world production of biomass of all kinds. The word "biomass" means living creatures, plants and animals and microbes, plus the organic remains that are left over when the creatures defecate or die.

Smil's Chapter 7 contains a comprehensive survey of the various kinds of biomass that drive the seasonal rhythms of the biosphere.

We do not know whether the increased yields observed in greenhouses with increased carbon dioxide are also occurring in open-air agriculture. Agricultural yields are limited by many factors other than carbon dioxide abundance. One factor that we know to be often limiting for plant growth is water abundance. If the supply of water is limiting, as it often is in times of drought, then increased carbon dioxide can still be helpful. The little pores in the leaves of plants have to be kept open for the plant to acquire carbon dioxide from the air, but the plant loses a hundred molecules of water through the pores for every one molecule of carbon dioxide that it gains. This means that increased carbon dioxide in the air allows the plant to partially close the pores and reduce the loss of water. In dry conditions, increased carbon dioxide becomes a water-saver and gives the plant a better chance to keep on growing.

The fundamental reason why carbon dioxide abundance in the atmosphere is critically important to biology is that there is so little of it. A field of corn growing in full sunlight in the middle of the day uses up all the carbon dioxide within a meter of the ground in about five minutes. If the air were not constantly stirred by convection currents and winds, the corn would not be able to grow. The total content of carbon dioxide in the atmosphere, if converted into biomass, would cover the surface of the continents to a depth of less than an inch. About a tenth of all the carbon dioxide in the atmosphere is actually converted into biomass every summer and given back to the atmosphere every fall. That is why the effects of fossil fuel–burning cannot be separated from the effects of plant growth and decay.

There are five reservoirs of carbon that are biologically accessible on a short timescale, not counting the carbonate rocks and the deep ocean, which are only accessible on a timescale of thousands of years. The five accessible reservoirs are the atmosphere, the land plants, the topsoil in which land plants grow, the surface layer of the ocean in which ocean plants grow, and our proved reserves of fossil fuels. The atmosphere is the smallest reservoir and the fossil fuels are the largest, but all five reservoirs are of comparable size. They all interact strongly with one another. To understand any of them, it is necessary to understand all of them. That is why planetary ecology is not an exact science like chemistry.

As an example of the way different reservoirs of carbon dioxide may interact with each other, consider the atmosphere and the topsoil. Greenhouse experiments show that many plants growing in an atmosphere enriched with carbon dioxide react by increasing their root-to-shoot ratio. This means that the plants put more of their growth into roots and less into stems and leaves. A change in this direction is to be expected, because the plants have to maintain a balance between the leaves collecting carbon from the air and the roots collecting mineral nutrients from the soil. The enriched atmosphere tilts the balance so that the plants need less leaf area and more root area. Now consider what happens to the roots and shoots when the growing season is over, when the leaves fall and the plants die. The new-grown biomass decays and is eaten by fungi or microbes. Some of it returns to the atmosphere and some of it is converted into topsoil.

On the average, more of the aboveground growth will return to the atmosphere and more of the belowground growth will become topsoil. So the plants with increased root-to-shoot ratio will cause an increased net transfer of carbon from the atmosphere into the topsoil. If the increase in atmospheric carbon dioxide due to fossil fuel–burning has caused an increase in the average root-to-shoot ratio of plants over large areas, then the possible effect on the topsoil reservoir will not be small. At present we have no way to measure or even to guess the size of this effect. The aggregate biomass of the topsoil of the United States is not a measurable quantity. But the fact that the topsoil is unmeasurable does not mean that it is unimportant.

Roughly speaking, half of the contiguous United States, not including Alaska and Hawaii, consists of mountains and deserts and parking lots and highways and buildings, and the other half is covered with plants and topsoil. Just to see how important an unmeasurable increase of topsoil may be, let us imagine that the increased root-to-shoot ratio of plants might cause an average net increase of topsoil biomass of one tenth of an inch per year over half the area of the contiguous United States. A simple calculation shows that the amount of carbon transferred from the atmosphere to the topsoil would be five billion tons per year. This amount is considerably more than the measured four-billion-ton annual increase of carbon in carbon dioxide in the atmosphere. So the increase of carbon dioxide in the atmosphere over the entire earth could be canceled out by an increase of topsoil biomass of a tenth of an inch per year over half of the contiguous United States.

A tenth-of-an-inch-per-year increase of topsoil would be exceedingly difficult to measure. At present we do not even know whether the topsoil of the United States is increasing or decreasing. Over the rest of the world, because of large-scale deforestation and erosion, the topsoil reservoir is probably decreasing. We do not know whether intelligent land management could ensure a growth of the topsoil reservoir by four billion tons of carbon per year, the amount needed to stop the increase of carbon dioxide in the atmosphere. All that we can say for certain is that this is a theoretical possibility and ought to be seriously explored.

Another problem mentioned by Smil that has to be taken seriously is a slow rise of sea level, which could become catastrophic if it continues to accelerate. We have accurate measurements of sea level going back two hundred years. We observe a steady rise from 1800 to the present, with an acceleration during the last fifty years. It is widely believed that the recent acceleration is due to human activities, since it coincides in time with the rapid increase of carbon dioxide in the atmosphere. But the rise from 1800 to 1900 was probably not due to human activities. The scale of industrial activities in the nineteenth century was not large enough to have had measurable global effects. A large part of the observed rise in sea level must have other causes. One possible cause is a slow readjustment of the shape of the earth to the disappearance of the northern ice sheets at the end of the ice age 12,000 years ago. Another possible cause is the large-scale melting of glaciers, which also began long before human influences on climate became significant. Once again, we have an environmental danger whose magnitude cannot be predicted until we know much more about its causes.

The most alarming possible cause of sea-level rise is the rapid disintegration of the West Antarctic ice sheet, which is the part of Antarctica where the bottom of the ice is far below sea level. Warming seas around the edge of Antarctica might erode the ice cap from below and cause it to collapse into the ocean. If the whole of West Antarctica disintegrated rapidly, sea level would rise by five meters, with disastrous effects on billions of people. However, recent measurements of the icecap show that it is not losing volume fast enough to make a significant contribution to the presently observed sea-level rise. It appears that the warming seas around Antarctica are causing an increase in snowfall over the icecap, and the increased snowfall on top roughly cancels out the decrease of ice volume caused by erosion at the edges. This is another situation in which we do not know how much of the environmental change is due to human activities and how much to long-term natural processes over which we have no control.

Another environmental danger that is even more poorly understood is the possible coming of a new ice age. A new ice age would mean the burial of half of North America and half of Europe under massive ice sheets. We know that there is a natural cycle that has been operating for the last 800,000 years. The length of the cycle is 100,000 years. In each 100,000-year period, there is an ice age that lasts about 90,000 years and a warm interglacial period that lasts about 10,000 years. We are at present in a warm period that began 12,000 years ago, so the onset of the next ice age is overdue. If human activities were not disturbing the climate, a new ice age might begin at any time within the next couple of thousand years, or might already have begun. We do not know how to answer the most important question: Does our burning of fossil fuels make the onset of the next ice age more likely or less likely?

There are good arguments on both sides of this question. On the one side, we know that the level of carbon dioxide in the atmosphere was much lower during past ice ages than during warm periods, so it is reasonable to expect that an artificially high level of carbon dioxide might stop an ice age from beginning. On the other side, the oceanographer Wallace Broecker[3] has argued that the present warm climate in Europe depends on a circulation of ocean water, with the Gulf Stream flowing north on the ocean surface and bringing warmth to Europe, while a countercurrent of cold water flows south in the deep ocean. So a new ice age could begin whenever the cold, deep countercurrent is interrupted. The countercurrent could be interrupted when the cold surface water in the Arctic becomes less salty and fails to sink, and the water could become less salty when the warming climate increases the Arctic rainfall. Thus Broecker argues that a warm climate in the Arctic may paradoxically cause an ice age to begin. Since we are confronted with two plausible arguments leading to opposite conclusions, the only rational response is to admit our ignorance. Until the causes of ice ages are understood in detail, we cannot know whether the increase of carbon dioxide in the atmosphere is increasing or decreasing the danger.

---

3. W. S. Broecker, "Thermohaline Circulation, the Achilles Heel of Our Climate System: Will Man-Made CO_2 Upset the Current Balance?," *Science*, Vol. 278 (1997), pp. 1582–1588, cited by Smil.

The biosphere is the most complicated of all the things we humans have to deal with. The science of planetary ecology is still young and undeveloped. It is not surprising that honest and well-informed experts can disagree about facts. But beyond the disagreements about facts, there is another deeper disagreement about values. The disagreement about values may be described in an oversimplified way as a disagreement between naturalists and humanists. Naturalists believe that nature knows best. For them the highest value is respect for the natural order of things. Any gross human disruption of the natural environment is evil. Excessive burning of fossil fuels, and the consequent increase of atmospheric carbon dioxide, are unqualified evils.

Humanists believe that humans are an essential part of nature. Through human minds the biosphere has acquired the capacity to steer its own evolution, and we are now in charge. Humans have the right to reorganize nature so that humans and biosphere can survive and prosper together. For humanists, the highest value is intelligent coexistence between humans and nature. The greatest evils are war and poverty, underdevelopment and unemployment, disease and hunger, the miseries that deprive people of opportunities and limit their freedoms. As Bertolt Brecht wrote in *The Threepenny Opera*, "Feeding comes first, morality second." If people do not have enough to eat, we cannot expect them to put much effort into protecting the biosphere. In the long run, preservation of the biosphere will only be possible if people everywhere have a decent standard of living. The humanist ethic does not regard an increase of carbon dioxide in the atmosphere as evil, if the increase is associated with worldwide economic prosperity, and if the poorer half of humanity gets its fair share of the benefits.

Vernadsky, as Smil portrays him, was a humanist. He foresaw the gradual transformation of the biosphere into a noosphere. The word "noosphere," a sphere of mind, means a planetary ecology designed and maintained by human intelligence. He recognized that, as the noosphere comes into existence, "the aerial envelope of the land as well as all its natural waters are changed both physically and chemically." He understood that the maintenance of a noosphere places heavy responsibilities on human shoulders. But he had faith in the ability of humans to rise to the challenge. The main conclusion of Vernadsky's thinking, and the main conclusion of Smil's book, is that life is complicated and any theory that attempts to describe its behavior in simple terms is likely to be wrong.

Postscript, 2006

After this review appeared, Vaclav Smil published another book, *Energy at the Crossroads: Global Perspectives and Uncertainties* (MIT Press, 2003), dealing directly with the practical issues of energy supply and demand. The new book makes a good complement to *The Earth's Biosphere*, which describes the larger framework of ecology within which practical policies must fit. I am grateful to Smil for sending me the new book, and sorry that I had not seen it when I wrote the review.

6

WITNESS TO A TRAGEDY

THOMAS LEVENSON IS a filmmaker who produces documentary films for public television. He has a sharp eye for the dramatic events and personal details that bring history to life. His book *Einstein in Berlin*[1] is a social history of Germany covering the twenty years from 1914 to 1933, the years when Albert Einstein lived in Berlin. The picture of the city's troubles comes into a clearer focus when it is viewed through Einstein's eyes. Einstein was a good witness, observing the life of the city in which he played an active role but remained always emotionally detached. He wrote frequent letters to his old friends in Switzerland and his new friends in Germany, recording events as they happened and describing his hopes and fears. His daily life and activities come intermittently into the narrative but are not the main theme. The main theme is the tragedy of World War I, a tragedy that began in 1914 but did not end in 1918. This tragedy continued to torment the citizens of Berlin through the years from 1918 to 1933 and led them finally to put their fate in the hands of Hitler. Hitler was able to gain his power over them because he promised to erase the tragedy and bring them back to the happy days of the empire when Germany was prosperous and united.

Every aspect of Einstein's life, the personal, the political, the

1. Random House, 2003.

scientific, and the philosophical, has been described in detail and analyzed in depth by his various biographers. The world does not need another Einstein biography. Fortunately, Levenson's book is not a biography. He has borrowed everything he needs from the published correspondence and the existing biographies of Einstein, with full acknowledgments and an excellent bibliography. The new and original aspect of this book is the context in which Einstein is placed. The context is a study in depth of the social pathology that gripped Berlin from the day Einstein arrived there in 1914 to the day he left in 1932.

The tragedy is a play in two acts, the first act being the years of war and the second act the years of the Weimar Republic. The most remarkable feature of the first act was the general belief among Einstein's friends in Berlin that the war was winnable. The war was widely welcomed as an opportunity for Germany to achieve its proper status as a great power. Einstein observed that his academic friends and colleagues were even more deluded with patriotic dreams of grandeur than the ordinary citizens that he met in the street. In a conversation with his Swiss friend Romain Rolland in 1915, he described how Berlin had gone to war. "The masses were immensely submissive, domesticated," he said. "The elites were worse. They were hungry, driven by their urge for power, their love of force, and the dream of conquest." As late as the summer of 1918, after the failure of the final German offensive on the western front, many of the leading German academics were still confident of victory.

The state of mind of the mandarins in Berlin was very different from the state of mind of their enemies in Paris and London. In Paris the war was seen as a desperate struggle for survival. The guns on the western front were close enough so that everyone in Paris could hear them. In Britain the war was seen as a tragedy that had done irreparable harm to Britain and to European civilization, no matter who won it. When the war came to an end in November 1918, the British public looked back on it as an unspeakable horror that should never under any circumstances be allowed to happen again. But a large part of the German public looked back on it differently, as a test of strength that they could have won if they had not been stabbed in the back by traitors at home. This book explains how that fatal German sense of betrayal came into being.

The second act of the tragedy is the story of the slow collapse of the Weimar Republic and the rapid rise of Hitler. Einstein was a firm supporter of the republic, but he saw which way the wind was blowing. One episode in the tragedy epitomizes the whole story. Erich Remarque's book *Im Westen Nichts Neues* was published in 1929 and immediately became an international best seller. It is the finest of all fictional accounts of World War I, seen through the eyes of a group of young Germans who die pointlessly in the carnage of the western front. In 1930 it was made into a Hollywood film, *All Quiet on the Western Front.* The film was shown all over the world, except in Germany. When the distributors of the film tried to show it in Berlin, Hitler's friend Joseph Goebbels organized a riot in the theater. Further Nazi demonstrations and violent protests against the film followed. And then the Weimar government banned the film throughout Germany. The Weimar authorities did not allow the German public to see the film because the Nazis considered it unpatriotic. This episode explains a mystery in my own family. One of my relatives is a lady, now ninety-four years old, who lived in Germany all her life and grew up in the Weimar years. Many years ago, I gave her Remarque's book to read and she found it very moving. "This book is wonderful," she said. "Why didn't they let us read it when it was published? That was before the Hitler time, but we were told that it was disgusting and shameful and respectable people should not read it." So the respectable Germans of her generation, even those who were not Nazis, did not read Remarque. I always wondered why, and now I know.[2]

2. The lady who did not read Remarque until it was too late was my mother-in-law, Gisela Jung. She died in March 2003. A few sentences of this review have been rewritten to avoid overlap with the review of Yuri Manin's *Mathematics and Physics* (Chapter 14).

II

War and Peace

7

BOMBS AND POTATOES

ON OCTOBER 16, 1945, General Leslie R. Groves presented J. Robert Oppenheimer with a certificate from the secretary of war, expressing the appreciation of the government for the work of the Los Alamos Laboratory. Oppenheimer responded with the following speech:

> It is with appreciation and gratitude that I accept from you this scroll for the Los Alamos Laboratory, for the men and women whose work and whose hearts have made it. It is our hope that in years to come we may look at this scroll, and all that it signifies, with pride.
>
> Today that pride must be tempered with a profound concern. If atomic bombs are to be added as new weapons to the arsenals of a warring world, or to the arsenals of nations preparing for war, then the time will come when mankind will curse the names of Los Alamos and Hiroshima.
>
> The peoples of this world must unite, or they will perish. This war, that has ravaged so much of the earth, has written these words. The atomic bomb has spelled them out for all men to understand. Other men have spoken them, in other times, of other wars, of other weapons. They have not prevailed. There are some, misled by a false sense of human history, who hold

> that they will not prevail today. It is not for us to believe that. By our works we are committed, committed to a world united, before this common peril, in law, and in humanity.

On the one side, these words of Oppenheimer. On the other side, memories of England in 1939. In 1939 in England, the younger generation was very sure that mankind must unite or perish. We had not the slightest confidence that anything worth preserving would survive the impending war against Hitler. The folk memory of England was dominated by the unendurable barbarities of World War I, and none of us could believe that World War II would be less brutal or less demoralizing. It was frequently predicted that as World War I had led to the collapse of society and the triumph of Bolshevism in Russia, so World War II would have the same effect in England.

When Neville Chamberlain declared war on Hitler in 1939, one of his first acts was to empty London hospitals of their patients. Chamberlain expected catastrophic air attacks to begin immediately; the hospitals were asked to be ready to handle 250,000 civilian casualties within the first two weeks, besides another 250,000 people who were expected to become permanently insane. These numbers were not based on fantasy; they were the estimates of military experts who extrapolated to the capability of the 1939 Luftwaffe the results achieved by much smaller forces in Spain and Ethiopia. The experts did not all agree on these numbers, but they agreed on the general order of magnitude. The public, fed by lurid newspaper and magazine articles, tended to view the approaching war in even more apocalyptic terms.

It was obvious to us young people in 1939 that war and surrender to Hitler were both unacceptable, both offering to us no substantial hope for the future. To escape from this dilemma, many of us took refuge in the gospel of Gandhi, believing that a nonviolent resistance to evil could defend our ideals without destroying them in the process.

The English pacifist movement of the late 1930s has not been kindly treated by history, but it was in fact neither cowardly nor muddle-headed. We made only one mistake; none of us in those days could imagine that England would survive six years of war against Hitler, achieve most of the political objectives for which the war had been fought, suffer only one third the casualties that we had had in World War I, avoid the massive and indiscriminate use of poison gas and biological weapons, and finally emerge into a world in which our moral and humane values were largely intact. When Chamberlain led us into war in 1939, his view of the outcome was probably as dark as ours, only he was sustained in his determination by the feeling that he had no honorable alternative.

I come at last to Tom Stonier's book *Nuclear Disaster*,[1] which is a thorough and straightforward study of the consequences of nuclear war. Stonier is a biologist, and this fact gives his analysis breadth which has been lacking in earlier studies by physical scientists. His conclusions are not quantitative but are clear and stark. He asserts that the United States would not survive in anything resembling its present form after a major thermonuclear exchange. He documents his conclusion with detailed discussion of the medical, ecological, and social problems of survival in a physically mutilated and contaminated country. He finds that although each problem by itself might well be overcome by energetic action and organization, all the problems together are likely to present insuperable difficulties. The life of the surviving postwar population is pictured as being "nasty, brutish and short" for many generations.

Stonier's knowledge of the physical and biological effects of nuclear explosions is solid and professional. His description of the economic and social effects is entirely plausible. Nevertheless, his total assessment of the long-range effect of nuclear war is necessarily

1. Meridian Books, 1963.

dependent on his personal judgment. Nobody can say for sure whether a population subjected to unprecedented horrors and privations would respond with apathetic despair or with heroic discipline. The problem here is to predict the psychological, moral, and spiritual reactions of people in circumstances for which we have no valid historical parallel.

Stonier describes at some length the reactions observed during and after the Irish potato famine of 1845–1848. This description is of absorbing interest, but its relevance to the problem of nuclear war is at best conjectural. In the end, readers of the book must decide for themselves, following their individual tastes or prejudices, whether they accept or reject Stonier's gloomy prognosis for the long-range recovery of civilization.

Just because the conclusions of Stonier's book depend so heavily on subjective judgment, it is important to view the book with a wider historical perspective. For this reason I began with Oppenheimer's speech and with the lessons of the 1930s. In the 1930s we held views about war very similar to those of Stonier, and these views turned out to be wrong. The experts who so grossly overestimated the effectiveness of bombing in 1939 made many technical errors, but their major mistake was a psychological one. They failed completely to foresee that the direct involvement of civilian populations in warfare would strengthen their spirit and social cohesion. The unexpected toughness and discipline of populations under attack was seen not only in England but even more strikingly in Germany, Japan, and the Soviet Union. Would the same qualities be shown in the United States after a nuclear attack? Stonier thinks not. I am not sure.

So we come back finally to the simple and profound words of Oppenheimer's speech. What we said about war in 1939 did not prevail. We learned in 1939–1945 that a war could still be fought and won without destroying the soul of a country. We learned that yielding to threats is the greater evil, and this is the lesson that most of us are now living by. When we in America apply this lesson to our dealings with the Soviet Union in the year 1964, are we misled by a false sense of human history? Is it a false sense of human history that teaches us that nationalism is still the strongest force in the world, stronger than the hydrogen bomb and stronger than humanity? These are some of the questions which Stonier's book does not answer.

Oppenheimer was certainly right in his basic perception, that history changed its course in 1945. Never again can a major war be fought in the style of World War II. And yet, international politics are being conducted on all sides as if the lessons of World War II still applied. History proceeds at its old slow pace, even if the course is changed. The transition from virulent nationalism to a world united must be stretched out over centuries. Meanwhile, we have to live in a precarious balance, between the apocalyptic warnings of Stonier on the one side and a possibly false sense of human history on the other. In spite of all uncertainties, it remains true that the catastrophes envisaged by Stonier may happen. It is well that we should be reminded of these dangers, and we must be grateful to Stonier for having reminded us of them, with his sober, thoughtful and eloquent book.[2]

---

2. This is the oldest piece in the collection, written in 1964. I have included it because the dilemma that it describes is still as real today as it was in 1964.

8

GENERALS

AT 2:30 PM ON August 31, 1946, the former chief of the Operations Staff of the German armed forces, Colonel-General Alfred Jodl, made his final statement to the Nuremberg War Crimes Tribunal:

> Mr. President and Justices of the court. It is my unshakable belief that history will later pronounce an objective and fair judgment on the senior military commanders and their subordinates. They, and with them the German armed forces as a whole, faced an insoluble problem, namely, to wage a war which they had not wanted, under a supreme commander who did not trust them and whom they only partially trusted, with methods which often contradicted their doctrines and their traditional beliefs, with troops and police forces not fully subject to their command, and with an intelligence service which was partly working for the enemy. And all this with the clear knowledge that the war would decide the existence or nonexistence of the beloved fatherland. They were not servants of Hell or of a criminal. They served their people and their fatherland.
>
> For myself, I believe that no man can do better than to struggle for the highest goal which he is in a position to achieve. That and nothing else was the guiding principle of my actions all

> along. And that is why, no matter what verdict you may pass on me, I shall leave this court with my head held as high as when I entered it many months ago. If anyone calls me a traitor to the honorable tradition of the German army, or if anyone says that I stayed at my post for reasons of personal ambition, I say he is a traitor to the truth.
>
> In this war, hundreds of thousands of women and children were destroyed by carpet-bombing, and partisans used without scruple whatever methods they found effective. In such a war, severe measures, even if they are questionable according to international law, are not crimes against morality and conscience. For I believe and profess: duty toward people and fatherland stands above every other. To do that duty was my honor and highest law. I am proud to have done it. May that duty be replaced in a happier future by an even higher one: duty toward humanity.

On October 10 he wrote a final letter to his friends in the German army:

> Dear friends and comrades. In the months of the Nuremberg trial I have borne witness for Germany, for her soldiers, and for history. The dead and the living crowded around me, giving me strength and courage. The verdict of the court went against me. That came as no surprise. The words which I heard from you were for me the true verdict. I was never proud in my life until now. Today I can and I will be proud. I thank you, and one day Germany will thank you, because you did not run away from one of her truest sons in his hour of need and death. Your future lives must not be filled with sadness and hate. Think of me only with respect and pride, just as you think of all the soldiers who died on the battlefields of this cruel war as they were required to do by law. Their lives were sacrificed to make Germany more

> powerful, but you should believe that they died to make Germany better. Hold fast to this belief and work for it all your lives.

On October 15 he wrote his last letter to his wife:

> And so I tell you at the end, you must live and overcome your grief. You must spread love around you and give help to those who need it. You must not make more of me than I was, or than I wanted to be. You must believe and make it known that I worked and fought for Germany and not for her politicians. Oh, I could go on writing like this forever, but now in my ears I hear the bugles playing taps, and the old familiar song—do you hear it, my love?—Soldiers must go home.

At 2 AM the next morning he was hanged.

Alfred Jodl was a Bavarian, not a Prussian. Probably that was one of the reasons Hitler chose him as chief of staff and kept him at his side throughout the war. Jodl nevertheless embodied the old Prussian tradition of military professionalism, with all its virtues and vices. Albert Speer, who sat with him in the dock at Nuremberg, wrote of him afterward: "Jodl's precise and sober defense was rather imposing. He seemed to be one of the few men who stood above the situation." For six years Jodl had worked, day after day and night after night, planning and organizing the campaigns in which millions died. He begged Hitler many times to relieve him of this responsibility, to give him a subordinate command at one of the fighting fronts. Hitler refused, and Jodl obeyed, steadfast up to the end. Jodl had sworn an oath, on his honor as a soldier, to obey Hitler as supreme commander. This oath of soldierly loyalty was for Jodl the unbreakable bond, holier than the Catholic faith in which he had been raised, stronger than his obligation to the welfare of the German people which he believed himself to be serving. On the day that he came to Berlin to take up his duties as chief of staff, a week before the German armies marched into Poland, he said to his wife: "This time I am afraid it looks like the real thing. I don't yet know for sure. But thank God, that is a problem for the politicians and not for us soldiers. I know only one thing; if we once get on board this boat there won't be any climbing out of it."

Jodl's personal religion and code of ethics were summed up in one word, *Soldatentum*, a German word which fortunately cannot be translated into English. The literal equivalent in English is "soldierliness," but the English word altogether misses the tone of solemnity which belongs to the German, and thereby misses the greater part of the meaning. The English word "militarism" means something else entirely. An accurate translation of *Soldatentum* would have to be a paraphrase: the profession of soldiering considered as a quasi-religious vocation. The emotional flavor of it is well conveyed in the writings of Jodl which I have quoted. In English, the word "chivalry" had once a similar aroma, but it became archaic and metaphorical after knights ceased to fight on horseback.

We are fortunate to have a biography of Jodl written by his widow, Luise.[1] Luise Jodl was a career woman, working with the same dedication as her husband in the bureaucracy of the general staff of the German armed forces. She shared her husband's code of honor and his professional pride. During the Nuremberg trial she helped the lawyers to prepare his defense. After he died, she wrote his biography while the events were still fresh in her mind. The book is valuable, not only for its authentic portrait of Alfred Jodl but also for its portrait of Luise. She, too, was a person of strong character and intelligence, driven to disaster by the ideal of *Soldatentum*. At the

1. *Jenseits des Endes: Leben u. Sterben des Generaloberst Alfred Jodl* (Beyond the End: Life and Death of Colonel-General Alfred Jodl) (Vienna: Molden, 1976). There is no English edition; the translations here are my own.

beginning of her book she placed a quotation from T. S. Eliot's poem "Little Gidding":

> *And what you thought you came for*
> *Is only a shell, a husk of meaning*
> *From which the purpose breaks only when*
> *it is fulfilled*
> *If at all. Either you had no purpose*
> *Or the purpose is beyond the end you figured*
> *And is altered in fulfilment.*

For the title of her book she took Eliot's words: "Beyond the End." Eliot wrote these words in the quietness of wartime England, in the early years of the war, when no end was in sight. The passage continues with lines which Luise Jodl must have known but did not choose to quote:

> *There are other places*
> *Which also are the world's end, some at*
> *the sea jaws,*
> *Or over a dark lake, in a desert or a city....*

One of those other places was Nuremberg, where Luise found herself in October 1946, alone among the ruins, faced with the tasks of piecing together the fragments of her husband's life and distilling some meaning from the dishonor of his death.

Perhaps the most brilliant field commander on either side in World War II was Hermann Balck. He commanded the motorized infantry regiment which led the decisive German breakthrough into France in 1940. Fighting later on the eastern front, he constantly surprised the Russians with unexpected moves and tactics. In the spring of 1945 he led the last German offensive of the war, holding off the Russian armies in Hungary long enough so that he could retreat in good order into Austria and finally surrender his troops to the Americans. He was, unlike Jodl, a real Prussian. He fought as Jodl was not permitted to fight, in the front lines with his soldiers. He was accused of no war crimes. In 1979, at the age of eighty-five, he entertained an American interviewer with his reminiscences.[2]

On Prussia:

> You need to see Prussia's situation in Europe, first of all. Prussia was a small country surrounded by superior forces. Therefore, we had to be more skillful and more swift than our enemies. That started perhaps with Frederick the Great at the battle of Leuthen where he defeated, and defeated thoroughly, a force of Austrians about twice as big as his own. In addition to being more clever than our opponents, we Prussians also needed to be able to mobilize much more quickly than our enemies.

On the breakthrough across the Meuse River in 1940:

> We knew in advance that we had to execute the crossing and I had already rehearsed it on the Moselle with my people. During this practice I had a couple of good ideas. First, every machine gun not occupied in the ground action was employed for air defense. Second, every man in the regiment was trained in the use of rubber boats. When we got to the Meuse, the engineers were supposed to be there, to put us across. They never arrived, but the rubber boats were there. So you see, if I hadn't trained my people, the Meuse crossing would have never happened.

2. Hermann Balck's reminiscences are in an interview with Pierre Sprey, published in two reports by Battelle Columbus Laboratories, Tactical Technology Center, Columbus, Ohio, January and April, 1979.

Which once again leads to the conclusion that the training of the infantryman can never be too many-sided....

The operation lay under intense French artillery fire. I had thrust forward to the Meuse with one battalion after some brief fights with the French outposts, and I had set up my regimental command post up front there on the Meuse, along with the forward battalion. I went along with them to make sure that some ass wouldn't suddenly decide to stop on the way. You know, the essence of the forward command idea is for the leader to be personally present at the critical place. Without that presence, it doesn't work.

On a tank battle in Russia in 1942:

I was heavily engaged in an attack with the 11th Panzer Division. Corps Headquarters called up at 7 o'clock in the evening and said that there had been a serious breakthrough 20 kilometers to my left, and that I should hurry over and take care of the breakthrough. I said, "Well, let me clean up the situation here and then I'll take care of the breakthrough." They said, "No, the situation on your left is terrible, and you've got to cease your attack immediately and clean up the breakthrough as fast as possible." I immediately gave the verbal order extricating us from the attack and directing the division to move and prepare for the new counterattack against the breakthrough 20 kilometers away. We launched our counterattack at 5 o'clock the next morning, and achieved such surprise that we bagged 75 Russian tanks without the loss of a single one of our own. Of course, one of the key reasons why we were able to achieve such quick movement was that I marched with the units. After all, the men were dead tired and nearly finished. I rode up and down the columns and asked the troops whether they preferred to march

> or bleed. To compare our speed with the Russians, I would estimate that a Russian armored division would have required at least 24 hours longer to have achieved the same movement we achieved in 10 hours. I had much less experience against the Americans, so I can only guess that the Americans would have been slightly faster than the Russians.

On attack and defense:

> It's quite remarkable that most people believe that attack costs more casualties. Don't even think about it; attack is the less costly operation.... The matter is, after all, mainly psychological. In attack, there are only 3 or 4 men in the division who carry the attack; all the others just follow behind. In defense, every man must hold his position alone. He doesn't see his neighbors; he just sees whether something is advancing towards him. He's often not equal to the task. That's why he's easily uprooted. Nothing incurs higher casualties than an unsuccessful defense. Therefore, attack wherever it is possible. Attack has one disadvantage; all troops and staffs are in movement and have to jump. That's quite tiring. In defense you can pick a foxhole and catch some sleep.

On generalship:

> There can be no fixed schemes. Every scheme, every pattern is wrong. No two situations are identical. That is why the study of military history can be extremely dangerous. Another principle that follows from this is: never do the same thing twice. Even if something works well for you once, by the second time the enemy will have adapted. So you have to think up something new. No one thinks of becoming a great painter simply by imitating Michelangelo. Similarly, you can't become a great

> military leader just by imitating so-and-so. It has to come from within. In the last analysis, military command is an art: one man can do it and most will never learn. After all, the world is not full of Raphaels either.

When Balck was a prisoner of war he resolutely refused to cooperate with American officers who asked him to contribute his reminiscences to an American historical project. Thirty years later, he had mellowed sufficiently to allow himself to be interviewed. The constant theme of his military career was learning to do more with less. He was always inventing new tricks to confound the enemy in front of him and the bureaucrats behind him. If I had to choose an epigraph for a biography of Balck, I would not take it from T. S. Eliot but from the old Anglo-Saxon poem commemorating the Battle of Maldon:

> *Thought shall be harder, heart the keener,*
> *Courage the greater, as our strength lessens.*

Balck, like the Saxons who fought the Danes at Maldon in the year 991, belonged to a tradition of soldiering older than *Soldatentum*, older than chivalry. Balck fought well because he enjoyed fighting well, and because he had a talent for it. As a professional soldier, he took his job seriously but not solemnly.

Jodl and Balck exemplify two styles of military professionalism, the heavy and the light, the tragic and the comic, the bureaucratic and the human. Jodl doggedly sat at his desk, translating Hitler's dreams of conquest into daily balance sheets of men and equipment. Balck gaily jumped out of one tight squeeze into another, taking good care of his soldiers and never losing his sense of humor. For Jodl, Hitler was Germany's fate, a superhuman force transcending right and wrong. Balck saw Hitler as he was, a powerful but not very competent politician. When Jodl disagreed with Hitler's plan to extend the German advance south of the Caucasus Mountains by dropping parachutists, the disagreement was for Jodl a soul-shattering experience. When Balck appealed directly to Hitler to straighten out a confusion in the supply of tanks and trucks, Hitler's failure to deal with the situation came as no surprise to Balck. "As it turned out," reports Balck, "Hitler never was able to gain control over the industry." Jodl went on fighting to the bitter end because he had made Hitler's will his highest law. Balck went on fighting because it never occurred to him to do anything else.

I chose my two examples of military professionalism from Germany because the German side of World War II displays the moral dilemmas of military professionalism with particular clarity. Both Jodl and Balck were good men working for a bad cause. Both of them used their professional skills to conquer and ravage half of Europe. Both of them continued to exercise their skills through the long years of retreat when the only result of their efforts was to prolong Europe's agony. Both of them appeared to be indifferent to the sufferings of the villagers whose homes their tanks were smashing and burning. And yet the judgment of Nuremberg made a distinction between them. Whether or not the Nuremberg tribunal was properly constituted according to international law, its decisions expressed the consensus of mankind at that moment of history. Jodl was hanged; Balck was set free; and the majority of interested bystanders agreed that justice had been done.

Roughly speaking, the distinction which the tribunal established and the public approved was a distinction between strategy and tactics. Balck was forgiven for waging war aggressively at the tactical level. Jodl was condemned for waging war aggressively at the strategic level. In the view of the tribunal, it is a sin for a soldier to plan campaigns for the overthrow and destruction of peaceful neighbors, but it is no sin for a soldier serving in such a campaign to be master of his trade. Rightly or wrongly, the public still approves the old tradition of military professionalism, giving honor and respect to soldiers who fight bravely in a bad cause.

The distinction between strategy and tactics is not the only difference between Jodl and Balck. There is another difference, which is equally important, although it was not used by the Nuremberg judges to justify their condemnation of Jodl, namely the distinction between soldiering and *Soldatentum*, the distinction between soldiering as a trade and soldiering as a cult. Balck was a likable character because he did not take himself too seriously. He went on winning battles, just as Picasso went on painting pictures, without pretentiousness or pious talk. He won battles because the skill came to him naturally. He never said that battle-winning was a particularly noble or virtuous activity; it was simply his trade. Jodl was unlikable and in the end diabolical because he set soldiering above humanity. He made his soldier's oath into a holy sacrament. He believed that he must be true to his ideal of *Soldatentum* even if it meant dragging Germany down to destruction. To be a good soldier was to him more important than to save what was left of Germany. He identified his duty as a soldier with loyalty to Hitler, and so he became infected with Hitler's insanity. The ideal of *Soldatentum* became an obsession, detached from reason, from reality, and from common sense.

Germany was an extreme case of military professionalism run wild, and the judgment at Nuremberg was an exceptional nemesis. But every country which gives an exalted status to its military leaders runs a risk of catching the German madness. There, but for the grace of God, go we. Something of the German madness infected the American South at the time of the Civil War. Like Germany, the states of the Confederacy made a cult of soldiering. It was no accident that the most brilliant generals of the Civil War were all fighting on the Southern side. Long before the war began, the Southern states had established a cultural tradition which encouraged their best minds to become professional soldiers. The tradition of exaggerated respect for military prowess was doubly disastrous for the South. It led to the overconfident enthusiasm and the illusions of military superiority with which the South went to war at the beginning. And it produced the spirit of sacrificial dedication in which the Southerners fought on to the bitter end, when the prolongation of the war was bringing to their country nothing but ruin and destruction. Robert Lee was a great general and a great gentleman, but all his tactical skill and strength of character only made the sufferings of his people heavier. The people of the Southern states, daily enduring death and destruction as a result of his activities, continued until the end to lavish upon him their unbounded love and admiration. When he returned to his home in Richmond after his surrender, he was greeted by an immense throng of cheering citizens. There was much in Lee that was noble and worthy of respect, but the hero-worship which surrounded him during and after the war was altogether disproportionate. The mystique of Lee distorted the Southerners' view of the world for a long time. Robert Lee was a greater general than Hermann Balck and a finer human being than Alfred Jodl, but his role in history was the same as theirs, to lead his beloved people to disaster. Any society which idolizes soldiers is tainted with a collective insanity and is likely in the end to come to grief.

England has been luckier in its choice of heroes. The British navy has been for hundreds of years the senior service, and the popular heroes of England have been admirals rather than generals. As a child in England, I learned to take it for granted that anybody making a career in the army must be intellectually subnormal. Army officers, on the stage or in real life, were figures of fun. Naval officers were also subject to occasional ridicule, but jokes about navy people were friendly rather than contemptuous. The navy commanded a certain respect even from the irreverent young. We were told that our Admiral Jellicoe, commander of the Grand Fleet during World War I, was the only man on either side who could have lost the war in one afternoon. He did not, after all, lose the war. Even though he was not very bright, and his one great battle, Jutland, was not a brilliant success, at least he did better than the generals on the western front. He kept cool and did not waste his ships in fruitless and unnecessary attacks.

The English language reflects the English bias in favor of sea captains. We have no word for soldierly virtue corresponding to the German *Soldatentum*. But we use naturally the word "seamanship," which has the same emotional resonance as *Soldatentum*, transferred from soldiers to sailors. Seamanship means not just technical competence in handling a ship; it includes also steadiness of nerve and strength of character, the virtues which Alfred Jodl subsumed under the heading *Soldatentum*. In England these virtues are perceived as belonging to sailors rather than to soldiers.

The English have not been exempt from the vice of military idolatry. A hundred years ago, Robert Louis Stevenson wrote an essay with the title "The English Admirals,"eloquently expressing the feelings of pride and glory which were then driving the worldwide expansion of the British Empire:

> Their sayings and doings stir English blood like the sound of a trumpet; and if the Indian Empire, the trade of London, and all the outward and visible ensigns of our greatness should pass away, we should still leave behind us a durable monument of what we were in these sayings and doings of the English Admirals. Duncan, lying off the Texel with his own flagship, the *Venerable*, and only one other vessel, heard that the whole Dutch fleet was putting to sea. He told Captain Hotham to anchor alongside. "I have taken the depth of the water," added he, "and when the *Venerable* goes down, my flag will still fly." And you observe that this is no naked Viking in a prehistoric period; but a Scotch member of Parliament, with a smattering of the classics, a telescope, a cocked hat of great size, and flannel underclothing. In the same spirit, Nelson went into Aboukir with six colours flying; so that even if five were shot away, it

should not be imagined he had struck.... And as our Admirals were full of heroic superstitions, and had a strutting and vain-glorious style of fight, so they discovered a startling eagerness for battle, and courted war like a mistress....

Trowbridge went ashore with the *Culloden*, and was able to take no part in the battle of the Nile. "The merits of that ship and her gallant captain," wrote Nelson to the Admiralty, "are too well known to benefit by anything I could say. Her misfortune was great in getting aground, while her more fortunate companions were in the full tide of happiness." This is a notable expression, and depicts the whole great-hearted, high-spoken stock of the English Admirals to a hair. It was to be "in the full tide of happiness" for Nelson to destroy five thousand five hundred and twenty-five of his fellow-creatures, and have his own scalp torn open by a piece of langridge shot. Hear him again at Copenhagen. A shot through the mainmast knocked the splinters about; and he observed to one of his officers with a smile, "It is warm work, and this may be the last to any of us at any moment," and then, stopping short at the gangway, added with emotion, "But, mark you, I would not be elsewhere for thousands..." The best artist is not the man who fixes his eye on posterity, but the one who loves the practice of his art. And instead of having a taste for being successful merchants and retiring at thirty, some people have a taste for high and what we call heroic forms of excitement. If the Admirals courted war like a mistress; if, as the drum beat to quarters, the sailors came gaily out of the forecastle, it is because a fight is a period of multiplied and intense experiences, and by Nelson's computation, worth thousands to anyone who has a heart under his jacket.[3]

---

3. R.L. Stevenson, "The English Admirals,"in *Virginibus Puerisque* (London: Chatto and Windus, 1897).

This is the stuff on which English children of the 1890s were raised. It is no better and no worse than the German patriotic literature of the same period. England and Germany were both in a mood of exuberant nationalism. Winston Churchill and Adolf Hitler grew up under its influence. Churchill and Hitler were both military romantics; both acted out their private dreams of glory by becoming great war leaders. And yet the effects of such glorification of the martial spirit in England and in Germany were vastly different. Churchill, and the English society which he represented, remained fundamentally sane, while Hitler's military visions led him and his society to paranoia and destruction. There were many historical and social reasons for the difference. No single factor by itself can explain such a profound divergence between neighboring cultures. But it may be that the most important cause of the difference between English and German destinies was the technical difference between the circumstances of sea and land warfare. War at sea, throughout the long period of British maritime ascendancy, was war for limited objectives. The technical limitations of sea power limited the human consequences of victory and defeat. None of Nelson's great victories resulted in the ruin of a province or the unconditional surrender of a country. The means of naval warfare determined the ends. Since the means were modest, the ends stopped short of insanity. No similar limitation of the means kept land warfare from escalating into wholesale conquest and genocide.

It is easy to go back into history and find other examples of sound and unsound military professionalism. Particularly illuminating in this respect is the contrast between the careers of Washington and Napoleon. Washington, fighting a war of limited objectives with means which were even more limited, laid the foundations for a durable and stable government in America. Everything which Washington built has lasted for two hundred years. Napoleon, fighting wars of unlimited objectives with armies greater than Europe had seen before, built an empire which crashed in ruins even before he was dead.

What can we learn from this picture gallery of soldiers, beginning with Jodl and Balck and Lee and ending with Washington and Napoleon? Professional soldiers and sailors have a necessary and honorable role to play in human affairs. The traditional respect which nations pay to military valor cannot be denied. As every country has a right to self-defense, every country has a right to give honor to its military leaders. But the honoring of military leaders brings deadly danger to mankind unless both the moral authority granted to them and the technical means at their disposal are strictly limited. Military power should never be confused with moral virtue, and military leaders should never be entrusted with weapons of unlimited destruction.

Nineteenth-century England was lucky to have military heroes who were modest both in their moral pretensions and in their material resources. Robert Louis Stevenson expressed in a nutshell the philosophy which allowed England to acquire an empire without losing a sense of proportion: "Almost everybody in our land, except humanitarians and a few persons whose youth has been depressed by exceptionally aesthetic surroundings, can understand and sympathize with an Admiral or a prize-fighter." This is the limit beyond which the pursuit of military glory should not go. A successful general or admiral should be honored no more and no less than a successful boxer.

The cult of military obedience and the cult of weapons of mass destruction are the two great follies of the modern age. The cult of obedience brought Germany to moral degradation and dismemberment. The cult of weapons of mass destruction threatens to bring us all to annihilation. It was, regrettably, the airmen of England who led the world into the cult of destruction. The Italian Giulio Douhet first preached the gospel of strategic bombing in the 1920s, but the British Sir Hugh Trenchard was the first to put Douhet's gospel into practice. England turned decisively away from the civilized nineteenth-century tradition of limited-objective warfare when Trenchard persuaded his government to build a force of heavy bombers with the deliberate aim of attacking the German civilian economy. The limited character of naval armament had made the exercise of British sea power in the nineteenth century peculiarly benign. The exercise of air power is subject to no such limitations. A strategy of strategic bombardment ensures that war will be total. And where England led the way into the era of strategic bombing, the United States was quick to follow. Already in the 1930s, England and America were set on the path which led to Hiroshima and Nagasaki. The cult of destruction possessed our bomber generals, and as a result of their activities we have been living under the threat of destruction ever since.

Every soldier who commands strategic forces, and every civilian strategist who theorizes about them, should from time to time imagine himself sitting in the dock at Nuremberg at the end of World War III and preparing his defense. Would his defense be any more convincing to the judges than the defense of Alfred Jodl? Jodl was, according to his own opinion and the opinion of his friends, an honorable man and a good soldier. Our strategic commanders and theorists are, so far as I know them, honorable men and good soldiers too. Jodl was condemned because his cult of obedience led to the death of millions. If our strategic commanders' cult of destruction should also lead to the death of millions, are they less deserving of condemnation?[4]

---

4. In this and the following two chapters, which were originally published in my book *Weapons and Hope* (Harper and Row) in 1984, I have omitted some passages which have been made obsolete by the collapse of the Soviet Union.

9

RUSSIANS

IOSIP SHKLOVSKY WAS a Russian astronomer of unusual brilliance, with several major discoveries to his credit. He was known to the Russian public as a writer of books and magazine articles describing the astronomical universe in a lively popular style. At scientific meetings he spiced his technical arguments with jokes and paradoxes. He had wide interests outside astronomy and could talk amusingly on almost any subject. He enjoyed unorthodox ideas, and he took a leading part in encouraging international efforts to listen for radio signals which might reveal the existence of intelligence in remote parts of the universe. In his professional life he projected an image of a happy, active, and successful man of the world. In private, like many Russian intellectuals, he was melancholic. He told me once that he had lived with a feeling of inner loneliness since he discovered, at the end of World War II, that he was the only one of his high school graduating class to have survived. He was the scientist in the class, and the authorities kept him out of the army to work on technical projects. The others went to the front and died. Russian citizens of Shklovsky's generation still bear the scars of war. Those who are younger grew up hearing tales of war told by their parents and grandparents. All alike carry deep in their consciousness a collective memory of suffering and irreparable loss. This is the central fact conditioning the Russian view of war. Russians, when they think of war, think of themselves not as warriors but as victims.

Another vignette of Russian life illustrates the same theme. It was a cold Sunday in late November, and I had the day free after a week of astronomical meetings in Moscow. The radio astronomer Nikolai Kardashev took me on a sightseeing trip to the ancient cities of Vladimir and Suzdal, halfway between Moscow and Gorky. We started before dawn and drove two hundred kilometers in darkness in order to arrive before the crowds. As we approached Suzdal we saw old monasteries shining golden in the light of the rising sun. Vladimir and Suzdal were places of refuge for monks and artists during the bitter centuries when Mongols and Tartars ruled in Russia. Both cities were taken and destroyed by the Mongols in 1238. They lay directly in the path of the army of Subutai, which swept across half of Europe in a merciless campaign of conquest. The inhabitants later rebuilt the cities, raised churches, and filled them with religious paintings. Vladimir and Suzdal lie far enough to the northeast so that they escaped the invasions which ravaged Kiev and Moscow in later centuries. Andrei Rublyov, the greatest painter of old Russia, worked at Vladimir in the fifteenth century. Buildings and paintings survive from the thirteenth century onward. Kardashev and I spent the day wandering from church to church among busloads of schoolchildren from Moscow and Gorky. The last stop on our tour was the city museum of Vladimir. Here we found the densest concentration of schoolchildren. The museum is in a tower over one of the ancient gates of the city. Its emphasis is historical rather than artistic. The main exhibit is an enormous diorama of the city as it was at the moment of its destruction in 1238, with every detail faithfully modeled in wood and clay. Across the plains come riding endless lines of Mongol horsemen slashing arms, legs, and heads off defenseless Russians whom they meet outside the city walls. The armed defenders of the city are on top of the walls, but the flaming arrows of the Mongols have set fire to the buildings behind them. Already a party of horsemen has broken into the city through a side gate and is beginning a general slaughter of the inhabitants. Blood is running in the streets and flames are rising from the churches. On the wall above this scene of horror there is a large notice for schoolchildren and other visitors to read. It says: "The heroic people of Vladimir chose to die rather than submit to the invader. By their self-sacrifice they saved Western Europe from suffering the same fate, and saved European civilization from extinction."

The diorama of Vladimir gives visible form to the dreams and fears which have molded the Russian people's perception of themselves and their place in history. Central to their dreams is the Mongol horde slicing through their country, swift and implacable. It is difficult for English-speaking people to share such dreams. The Russian experience of the Mongol invasions is so foreign to us that we gave the word "horde" a new and inappropriate meaning when we borrowed it into our language. English-speaking people came to Asia as traders and conquerors protected by a superior technology. Our view of Asia is mirrored in the image which the word "horde" conveys to an English-speaking mind. A horde in our language is a sprawling, undisciplined mob. In Russian and in the original Turkish, a horde is a camp or a tribe organized for war. The organization of the Mongol horde in the thirteenth century was technically far in advance of any other military system in the world. The Mongols could travel and maintain communications over vast distances; they could maneuver their armies with a speed and precision which no other power could match. It took the Russians 150 years to learn to fight them on equal terms, and three hundred years to defeat them decisively. The horde in the folk memory of Russia means an alien presence moving through the homeland, ravaging and consuming the substance of the people, subverting the loyalty of their leaders with blackmail and bribes. This is the image of Asia which three centuries of suffering implanted in the Russian mind. It is easy for us in the strategically inviolate West to dismiss Russian fears of China as "paranoid." If we had lived for three centuries at the mercy of the alien horsemen, we would be paranoid too.

British prime ministers, soon after they come into office, customarily visit Washington and Moscow to get acquainted with American and Russian leaders. When Prime Minister James Callahan made his state visit to Moscow he had two amicable meetings with Chairman Leonid Brezhnev. At the end of the second day he remarked that he was happy to discover that there were no urgent problems threatening to bring the United Kingdom and the Soviet Union into conflict. Brezhnev then replied with some emphatic words in Russian. Callahan's interpreter hesitated, and instead of translating Brezhnev's remark asked him to repeat it. Brezhnev repeated it and the interpreter translated: "Mr. Prime Minister, there is only one important question facing us, and that is the question whether the white race will survive." Callahan was so taken aback that he did not venture either to agree or to disagree with this sentiment. He made his exit without further comment. What he had heard was a distant echo of the Mongol hoofbeat still reverberating in Russian memory.

After the Mongols, invaders came to Russia from the West—from Poland, from Sweden, from France, and from Germany. Each of the invading armies was a horde in the Russian sense of the word, a disciplined force of warriors superior to the Russians in technology, in mobility, and in generalship. Especially the German horde invading Russia in 1941 conformed to the ancient pattern. But the Russians had made some progress in military organization between 1238 and 1941. It took them three hundred years to drive out the Mongols but only four years to drive out the Germans.

During the intervening centuries the Russians, while still thinking of themselves as victims, had become in fact a nation of warriors. In order to survive in a territory perennially exposed to invasion, they maintained great armies and gave serious study to the art of war. They imposed upon themselves a regime of rigid political unity and military discipline. They gave high honor and prestige to their soldiers, and devoted a large fraction of their resources to the production of weapons. Within a few years after 1941, the Russians who survived the German invasion had organized themselves into the most formidable army on earth. The more they think of themselves as victims, the more formidable they become.

Tolstoy's *War and Peace* is the classic statement of the Russian view of war. Tolstoy understood, perhaps more deeply than anyone else, the nature of war as Russia experienced it. He fought with the Russian army against the British and French at Sevastopol. He spent some of his happiest years as an artillery cadet on garrison duty in the Caucasus. In *War and Peace* he honored the courage and steadfastness of the ordinary Russian soldiers who defeated Napoleon in spite of the squabbles and blunders of their commanders. He drew from the campaign of 1812 the same lessons which a later generation of soldiers drew from the campaigns of World War II. He saw war as a desperate improvisation, in which nothing goes according to plan and the historical causes of victory and defeat remain incalculable.

Tolstoy's thoughts about war and victory are expressed by his hero Prince Andrei on the eve of the Battle of Borodino. Andrei is talking to his friend Pierre:

> "To my mind what is before us tomorrow is this: a hundred thousand Russian and a hundred thousand French troops have met to fight, and the fact is that these two hundred thousand men will fight, and the side that fights most desperately and spares itself least will conquer. And if you like, I'll tell you that whatever happens, and whatever mess they make up yonder, we shall win the battle tomorrow; whatever happens we shall win the victory." "So you think the battle tomorrow will be a

> victory," said Pierre. "Yes, yes," said Prince Andrei absently. "There's one thing I would do, if I were in power," he began again, "I wouldn't take prisoners. What sense is there in taking prisoners? That's chivalry. The French have destroyed my home and are coming to destroy Moscow; they have outraged and are outraging me at every second. They are my enemies, they are all criminals to my way of thinking.... They must be put to death.... War is not a polite recreation, but the vilest thing in life, and we ought to understand that and not play at war. We ought to accept it sternly and solemnly as a fearful necessity."

The battle was duly fought, and Prince Andrei was mortally wounded. The Russians lost, according to the generally accepted meaning of the word "lose": half of the Russian army was destroyed; after the battle the Russians retreated and the French advanced. And yet, in the long view, Prince Andrei was right. Russia's defeat at Borodino was a strategic victory. Napoleon's army was so mauled that it had no stomach for another such battle. Napoleon advanced to Moscow, stayed there for five weeks waiting for the Tsar to sue for peace, and then fled with his disintegrating army in its disastrous stampede to the West. "Napoleon," concludes Tolstoy, "is represented to us as the leader in all this movement, just as the figurehead in the prow of a ship to the savage seems the force that guides the ship on its course. Napoleon in his activity all this time was like a child, sitting in a carriage, pulling the straps within it, and fancying he is moving it along."[1]

I have been lucky to have had as a friend and colleague for forty years George Kennan, who spent the first half of his life as a diplomat serving in Russia and other countries of Eastern Europe, and the second half as a historian at the Institute for Advanced Study in Princeton. He died in 2005 at the age of 101. His fate, throughout his long

1. Leo Tolstoy, *War and Peace*, translated by Constance Garnett (Random House, 1931).

double career as government servant and independent scholar, was to be a teller of complicated truths to people who prefer simple illusions. Sometimes he has been driven close to despair by the inability of the American political system to pay attention to people with expert knowledge of the world outside. In 1944, when he was a member of the American embassy staff in Moscow, deeply troubled by the gap between American perception and Russian realities, he wrote a thirty-five-page essay summarizing his firsthand view of Russia, for the benefit of his superiors in Washington. Kennan afterward described the concluding sentences of this essay as "a melancholy, but for me personally most prophetic, passage":

> There will be much talk about the necessity for "understanding Russia"; but there will be no place for the American who is really willing to undertake this disturbing task. The apprehension of what is valid in the Russian world is unsettling and displeasing to the American mind. He who would undertake this apprehension will not find his satisfaction in the achievement of anything practical for his people, still less in any official or public appreciation for his efforts. The best he can look forward to is the lonely pleasure of one who stands at long last on a chilly and inhospitable mountaintop where few have been before, where few can follow, and where few will consent to believe that he has been.

For sixty years after those words were written, Kennan continued, as diplomat and historian, to bring us his reports from the mountaintop. At the end he knew that his efforts had not always been unappreciated.

My own view of the mountaintop has been derived partly from conversations with Kennan, partly from brief scientific visits to Russia, and partly from readings in Russian literature. The readings in

Russian literature began earliest and have left the deepest impression. As a teenager I worked my way through *The Oxford Book of Russian Verse*, guided by Maurice Baring's magnificent introduction. I found there, among other things, the poem "On the Field of Kulikovo" by Alexander Blok, which tells more about the Russian view of war than a whole library of strategic analysis. The Battle of Kulikovo was fought a century and a half after the destruction of Vladimir, but Blok's poem carries the same haunting message as the diorama in the Vladimir museum. Blok is riding over the steppe with the Russian horsemen the night before the battle:

I am not the first, nor the last, warrior,
Many years more will my country suffer.

At Kulikovo the Russians for the first time defeated a Tartar horde. The battle was a turning point in the centuries-long struggle of Russians against Tartars, not the end of the struggle but a new beginning. Blok wrote his poem in the year 1908, at a time of peace and prosperity, but he felt already the shadow of approaching storms:

I perceive you now, beginning
Of high turbulent days. Once more
Over the enemy camp the winging
Of swans is heard, swans trumpeting War.[2]

Ten years later, in January 1918, three months after the Bolshevik seizure of power, amid the chaos and cold of revolutionary Petrograd, Blok wrote his greatest poem, "The Twelve," which told us more about the nature of Soviet power than a whole library of Kremlinology. The twelve are a group of young soldiers of the Red Guard,

2. *The Oxford Book of Russian Verse*, edited by Maurice Baring (Clarendon Press, 1924).

marching through the city in a snowstorm, rough and tough and profane and trigger-happy:

"Grip your gun like a man, brother.
We'll pump some lead into Holy Russia,
Ancient, peasant-ridden, fat-arsed Mother
Russia.
Freedom, Freedom! Down with the Cross!"
"Open your cellars: quick, run down!
The scum of the earth are hitting the town!"
Abusing God's name as they go,
The twelve march onward through the snow,
Prepared for anything,
Regretting nothing.

In the final scene of the poem, the twelve are chasing a shadowy figure who lurks in a snowdrifted alleyway. They shout at the figure to surrender, then open fire. The echoes of their gunshots die away. The howling of the storm continues:

So they march with sovereign tread.
Behind them limps the hungry dog,
Ahead of them, carrying the blood-red flag,
Unseen in the blizzard,
Untouched by the bullets,
Stepping soft-footed over the snow
In a swirl of pearly snowflakes,
Crowned with a wreath of white roses,
Ahead of them goes Jesus Christ.[3]

---

3. Alexander Blok, *The Twelve and Other Poems*, translated by John Stallworthy and Peter France (Oxford University Press, 1970).

Blok remained in Russia until his death in 1921. Shortly before he died, he reaffirmed the vision which "The Twelve" had recorded:

> I do not go back on what I wrote then, because it was written in harmony with the elemental: for instance, during and after the writing of "The Twelve," for some days I physically felt and heard a great roar surrounding me, a continuous roar, probably the roar of the collapse of the old world.... The poem was written in that exceptional and always very brief period when the passing revolutionary cyclone raises a storm in every sea.... The seas of nature, life and art were raging and the foam rose up in a rainbow over them. I was looking at that rainbow when I wrote "The Twelve."

My own personal encounter with the armed forces of the Soviet revolution occurred in a later and more tranquil time. It was in May 1956, when the Russians organized the first postwar international meeting of high-energy physicists in Moscow. Russian experimental work in high-energy physics had previously been kept secret, for reasons which had little to do with military security. The last years of Stalin's life had been years of terror and silence for Russian intellectuals; even in the nonpolitical domain of physical science, publication had been severely restricted and contacts with foreign scientists almost nonexistent. When Stalin died, the icy grip of secrecy slowly weakened. In 1954 Ilya Ehrenburg was allowed to publish his novel *The Thaw*, which described the fresh stirrings of Russian life after the long winter. By 1956 the physicists were ready to celebrate the return of spring with a big conference to which colleagues from all over the world were invited. The conference was a joyful occasion for the Russians and for us too. Old friendships were renewed and new friendships established. The Russian newspapers gave us front-page coverage, and proudly described how the great leaders of international science were now coming to Moscow to learn about the great achievements of Soviet scientists.

After the Moscow meeting ended, I went with a group of foreign scientists to Leningrad. Accompanied by two Intourist guides, we went sightseeing along the shore to the west of the city. We walked by mistake into a coast guard station, evidently a restricted military area. An ordinary Russian seaman came out to shoo us away, shouting *Nelzya*, which means "forbidden." At that moment we noticed that our guides, afraid of being held responsible for our error, were walking rapidly away in the opposite direction. So we stayed and had a friendly chat with the seaman in our broken Russian. When I said we were foreign scientists, he broke out into a broad smile and said, "Oh, I know who you are. You are the people who came to the meeting in Moscow, and you know all about pi-mesons and mu-mesons." He pulled out of his pocket a crumpled copy of *Pravda* which contained a report of our proceedings. After that, he invited us into the station and proudly introduced us to his comrades. We sat with them for some minutes and did our best to explain to them what we had learned in Moscow about pi-mesons and mu-mesons. When we said good-bye, our host shook our hands warmly and said, "Why do you not come to our country more often? Be sure to tell the people in your countries, and your wives and children, that we would like to see more of them." As I walked back into Leningrad and reflected upon this encounter, I found myself sadly wondering whether an average American coast guard sentry, confronted unexpectedly with a group of Russian physicists speaking broken English, would have greeted them with equal friendliness and understanding.

Postscript, 2006

My most recent trip to Russia was in October 2003. Another unexpected encounter, this time with a young woman tour guide at the monastery of Sergiev Posad north of Moscow. Instead of talking about the churches and the famous works of art, she talked about her own spiritual experiences as a believer. She told how her life was changed when she came to one of the old tombs and smelled a holy scent emerging from the tomb. She knew then that she was called to be a guide and teach others about the mystical powers of the old saints. This is a religion very different from our Western versions of Christianity. It is a religion based on holy magic rather than Bible stories, mystical dreams rather than theological arguments.

10

PACIFISTS

THOUGHTLESSLY I SAID to the Russian sailor in the coast guard station, "You should also come and see us in America." He looked at me, laughing, with his broad young face. "How could we come to America? That's impossible. We are warriors." It was strange to hear him use that word, *voyenniye*—"warriors." He looked so unwarlike, sitting with his friends around the table and chatting with us about pi-mesons and mu-mesons. And yet the word spoke truth. His trade was war. He belonged to that ancient brotherhood of warriors which Alexander Blok described in his poems, the horsemen riding by night over the field of Kulikovo, the twelve marching in the snowstorm through the desolate streets of Petrograd. All his friendliness, his intellectual curiosity, his boyish humor could not alter the fact that he was a willing tool of Soviet power. A warrior he was, and a warrior he would remain, even after he finished his term of military service and found his niche in civilian society. All his life, he would be proud to have been a part of the Soviet navy. If ever he was called to sail into battle and die for his country, he would hesitate no more than those who sailed with Nelson at Trafalgar. If ever he was called to launch the missile that would obliterate a city, he would hesitate no more than those who aimed the bombs at Hiroshima and Nagasaki. When I imagine nuclear war, the nightmare begins with that young Russian sailor pressing the button which blows us all to smithereens, and as he presses it, he says, "We are warriors," with that same laughing voice of murderous innocence which I heard in Leningrad long ago.

Is there no other way? Is there no other tradition for our young men to follow than the tradition of warriors marching into battle to defend the honor of their tribe? Indeed there is another tradition, the tradition of pacifism, which also has a long and honorable history. For hundreds of years there have been religious sects which held warfare to be contrary to the will of God. Anabaptists and Quakers were preaching the gospel of nonviolence in the seventeenth century, and suffering persecution for their beliefs. This old tradition of nonviolence was personal rather than political. The Quakers allowed no authority to come between the individual conscience and God. They refused, as individuals, to bear arms or to take any part in the waging of wars. They did not seek political power for themselves or attempt to control the actions of governments. They simply declared that they would not take any action forbidden by their consciences. The tradition of personal pacifism which they established has proved durable. It has lasted for three hundred years and has taken root in many countries. Pacifism as a code of personal ethics has proved itself able to weather the storms of war and political change.

Pacifism as a political program is a more recent development. A political pacifist is one who advocates the ethic of nonviolence as a program for a political movement or for a government. Theorists of pacifism make a sharp distinction between personal and political pacifism. In the real world, this distinction is useful but never sharp. There is a continuum of pacifism, extending all the way from the private faith of the traditional conscientious objector to the modern rituals of nonviolent demonstration staged by political action groups in front of television cameras. Pacifism may be a matter of individual conscience or a matter of tactical calculation. Most commonly it is a mixture of both. If pacifism is ever to prevail in the modern world, it must be both personal and political, cherishing the deep roots of the religious pacifist tradition and at the same time exploiting the opportunities provided by modern communications to mobilize public protest. Gandhi, the first and greatest of modern political pacifists, showed us how it can be done.

The Quakers stand in the middle of the pacifist continuum, not so fully engaged in politics as Gandhi, not so detached as the Amish of Pennsylvania, who try to withdraw altogether from the violence and evil of the world. Quakers live in the world of anger and power and seek to mitigate its evils. The Quaker ethic has always encouraged its adherents to concern themselves with other people's sufferings. "Concern" in the Quaker vocabulary means more than sympathy; it means practical help for people in need and practical intervention against injustice. Large numbers of Quakers, following the example of their founder, George Fox, express their concern by campaigning in the political arena for humanitarian and pacifist ideals. But they act as individuals, not as an organized movement. Perhaps the main reason for the durability of the Quakers' influence is the fact that they are tied to no government and no party. Their pacifism is a private commitment based on conscience, not a political tactic dependent on success or popularity. They are not, like the followers of Gandhi, liable to defect from their pacifist principles when the political winds change.

The great and permanent achievement of the Quakers was the abolition of slavery. This social revolution, with the accompanying profound changes in public morality, took centuries to complete and was not the work of Quakers alone. But the earliest agitators against slavery were mostly Quakers. All through the eighteenth century, in England and in America, Quakers were prime movers in the uphill struggle, first to put an end to the profitable trade in fresh slaves from Africa, and later to put an end to the profitable exploitation of slaves wherever they happened to be. My great-great-great-uncle Robert Haynes was a prominent citizen of the island of Barbados, owner of several sugar plantations and several hundred slaves. In his diary for the year 1804 he complained bitterly of the public agitation against slavery which was then gathering strength in England. He knew who his enemies were. "I am likewise minded," he wrote, "to attribute a fair share of the blame to the underhand activities of a sect known as Quakers. These, from the very beginnings of the settlement of our island having played a very subtle—and in these days all too little heeded—part in the instigation of others to rebellion, at the same time openly avowing their detestation to any form of violence! Not scrupling, withall, to avail themselves fully of the safety and protection afforded them by the laws and defenses of this country. All this savouring of cant and hypocrisy such as I, for one, find hard to stomach."

The next item in Haynes's diary explains the violence of his feelings. "Attempted rising of slaves in some parts of the Island. The above quickly suppressed—the immediate shewing of discipline taking excellent and speedy effect—but at the same time a general anxiety thus engendered by no means, even now, wholly allayed."[1] Four years later the British Parliament passed the act which put an end to the slave trade, with effective criminal penalties. Haynes continued for twenty-five years longer to enjoy an uneasy dominion over his slaves on the island. But he lived long enough to see the Quakers finally victorious, his slaves freed, and the old order of society on the island overthrown. Handsomely compensated with a cash payment for his slaves by the Act of Parliament of 1833, he moved to England and lived the rest of his life at Reading in comfortable retirement.

What were the ingredients of the Quakers' success? First of all, moral conviction. They never had any doubt that slavery was a moral evil which they were called upon to oppose. Second, patience. They continued their work, decade after decade, undiscouraged by setbacks

1. *The Barbadian Diary of General Robert Haynes, 1787–1836*, edited by Everil M. W. Cracknell (Medstead: Azania Press, 1934).

and failures. Third, objectivity. A large part of their work consisted of careful collection of facts and statistics which both sides in the dispute came to accept as accurate. It was the fact-gathering activities of the Quakers in Barbados which particularly infuriated my great-great-great-uncle. Fourth, willingness to compromise. The Quakers were concerned to free the slaves, not to punish the slave owners. They accepted the fact that slaves were an economic asset and that the owners were entitled to fair compensation for the loss of their property. The slave owners were not to be humiliated. As a result, even my great-great-great-uncle in the end swallowed his pride and quietly pocketed his cash settlement. The willingness of the British abolitionists to buy out the slave owners made the crucial difference between the peaceful liberation of the West Indian slaves in 1833 and the bloody liberation of the American slaves thirty years later. The British government paid the slave owners twenty million pounds. The cost of the American Civil War was considerably higher.

The abolition of nuclear weapons is a task of the same magnitude as the abolition of slavery. Nuclear weapons are now, as slavery was two hundred years ago, a manifestly evil institution deeply embedded in the structure of our society. Most people nowadays, if they think about nuclear weapons at all, worry about nuclear bombs in the hands of terrorists. They imagine terrorists carrying one or two nuclear bombs in cars or trucks and exploding them in New York or Washington. One or two nuclear bombs exploding in a city would be a disaster much greater than the destruction of the World Trade Center in 2001. People are right to worry about terrorist bombs. But they ought to worry much more about the thousands of nuclear weapons that are not in the hands of terrorists but in the hands of national governments. Terrorist bombs could kill millions of people, while national nuclear weapons could kill hundreds of millions. National nuclear weapons used in a major war could destroy whole countries, including our own. And since the United States maintains the largest and most powerful deployments of nuclear weapons, we carry the largest share of moral responsibility for their continued existence.

People who hope to push the fight for the abolition of war to a successful conclusion must bring to their task the same qualities which won the fight for the abolition of slavery: moral conviction, patience, objectivity, and willingness to compromise. Those who fought against slavery two hundred years ago made a historic compromise which opened the way to their victory; they decided to concentrate their efforts upon the prohibition of the slave trade and to leave the total abolition of slavery to their successors in another generation. They saw that the slave trade was a more glaring evil than slavery itself and more vulnerable to political attack. They were able to mobilize against the slave trade a coalition of moral and economic interests which could not at that time have been brought together in the cause of total abolition. There is a lesson here for the peace movements of today. The ultimate aim of peace movements is the total abolition of war. All war is evil, but the use of nuclear weapons is a more glaring evil, and the abolition of nuclear weapons is a more practical political objective than the abolition of war. Modern pacifists, like the Quakers of the eighteenth century, would be well advised to attack the more vulnerable evil first. After we have succeeded in abolishing nuclear weapons, the abolition of war may become a feasible objective for later generations, but from here it is out of sight.

Pacifism as a political cause has suffered from the fact that its greatest leaders have been men of genius. People of outstanding genius, transcending the beliefs and loyalties of the tribe in which they happen to be born, tend naturally toward pacifism. Unfortunately, people of genius do not usually make good politicians. Gandhi was one of the rare exceptions. Genius and the art of political compromise do not sit easily together. Except for Gandhi, the great historic figures of pacifism have been prophets rather than politicians. Jesus in Judea, Tolstoy in Russia, Einstein in Germany, each in turn has set for mankind a higher standard than political movements can follow.

When Tolstoy wrote *War and Peace*, he was a Russian patriot, sympathetic to the martial spirit of his soldier characters and proud of their bravery. His skeptical realism belongs squarely, as Alexander Blok's fevered romanticism does not, in the mainstream of Russian patriotic literature. But Russian patriotism was too narrow a frame for Tolstoy's genius. At the age of fifty he experienced a religious conversion to the gospel of peace. He repudiated the sovereignty of all national governments, including his own. He cut himself off from the aristocratic society in which he had formerly lived. And for the last thirty years of his life he preached the ethic of nonviolence in its most uncompromising form. He demanded that we not only refuse to serve in armies and navies but also refuse to cooperate in any way with coercive activities of governments. Revolutionary action against governments was forbidden too; those who oppose a government with violence cannot lead the way to the abolition of violence. He called us to follow a way of life based on strict obedience to the words of Jesus: "Ye have heard that it hath been said, an eye for an eye and a tooth for a tooth: but I say unto you, that ye resist not evil; but whosoever shall smite thee on the right cheek, turn to him the other also."

The Tsar's government was wise enough not to lay hands on Tolstoy or to attempt to silence him. Only the young men who followed his teaching and refused military service were put in prison or exiled to Siberia. Tolstoy himself lived unmolested on his estate at Yasnaya Polyana with his faithful disciples and his disapproving wife. He corresponded with the young Gandhi. He became a prophet and spiritual leader for pacifists all over the world. Wherever he saw cruelty and oppression, he spoke out for the victims against the oppressors. He warned the wealthy and powerful in no uncertain terms of the explosion of violence to which their selfishness was leading: "Only one thing is left for those who do not wish to change their way of life, and that is to hope that things will last my time—after that, let happen what may. That is what the blind crowd of the rich are doing, but the danger is ever growing and the terrible catastrophe draws nearer." The wealthy and powerful listened politely to his warning and continued on the course which led to the cataclysms of 1914 and 1917. The situation of Tolstoy at the end of his life was similar to the situation of Einstein fifty years later, the venerable white-bearded figure, wearing a peasant blouse as a symbol of his contempt for rank and privilege, universally respected as a writer of genius, disdained by practical politicians as a cantankerous old fool, loved and admired by the multitude as spokesman for the conscience of mankind.

A hundred years have now passed since Tolstoy's conversion, and the power of nationalism over men's minds is as strong as ever. There was perhaps a chance, at the end of the nineteenth century and the beginning of the twentieth, that the working people of Europe would unite in a common determination not to be used as cannon fodder in their masters' quarrels. This was the dream which Tolstoy dreamed, and it was shared by many of the leaders of workers' organizations in various European countries during the years before 1914 when these organizations were growing rapidly in membership and power. The dream was an international brotherhood of workers united in loyalty to socialist and pacifist principles. The dream was an international general strike that would become effective on the day of declaration of war and would leave the generals of the belligerent armies without soldiers to command. Among the leaders who believed in international brotherhood as a practical political program for the workers of the world, Jean Jaurès of France was outstanding. Jaurès was an experienced politician, representing the French Socialist Party in the Chamber of Deputies, and reelected repeatedly by his constituency of miners. He was a patriotic Frenchman and never advocated unilateral disarmament or unconditional pacifism. He knew personally the Socialist leaders in Germany and Austria and understood the ambiguities of their position. But he believed with passionate conviction in the possibility that an international general strike against war could be successful. This dream collapsed on July 31, 1914, when the German, Austrian, and Russian armies were already mobilizing for war, the workers in each country had forgotten their international brotherhood and were marching obediently to the frontiers to defend their respective fatherlands, and Jaurès, sitting disconsolate at his supper in a restaurant in Paris, was shot dead by a fanatical French patriot.

Tolstoy's radical pacifism never became a serious political force in Europe, and least of all in Russia, either before or after the revolution. The only effective action of workers against war occurred in 1917, when Lenin encouraged the soldiers of Alexander Kerensky's government to desert from the front lines where they were fighting the Germans. But this desertion was not the fulfillment of Jaurès's dream of an international strike against war; it was merely an opening move in the new war for which Lenin was preparing. As soon as Lenin had seized power, he organized a new army and used it to defend his territory against the remnants of the old army in the civil war of 1918–1921. Neither the Tsar before the revolution nor Lenin afterward hesitated to spill blood; neither had difficulty in finding an ample supply of young Russians willing to kill or to die for the defense of Russia against her enemies. The seeds of Tolstoy's gospel of nonviolence fell mostly upon stony soil as they were carried all over the world, and nowhere was the soil stonier than in his native Russia.

The great blossoming of nonviolence as a mass political movement was the work of Gandhi in India. For thirty years he led the fight for Indian independence and held his followers to a Tolstoyan code of behavior. He proved that *satyagraha*, soul-force, can be an effective substitute for bombs and bullets in the liberation of a people. *Satyagraha*, a word and a concept invented by Gandhi, means much more than nonviolence. *Satyagraha* is not merely passive resistance or abstention from violent actions. *Satyagraha* is the active use of moral pressure as a weapon for the achievement of social and political goals. Gandhi used *satyagraha* impartially to castigate the British governors of India and his own followers, whenever they strayed from the path of nonviolence. With his Hindu background and his London lawyer's training, he understood the psychology of Indian peasants and of imperial government officials, and succeeded in bending them both to his will. The chief tools of *satyagraha* were civil disobedience, the peaceful but ostentatious breaking of laws imposed by the alien authorities, and the fast unto death, a personal hunger strike in which Gandhi repeatedly wagered his life in order to compel friends and enemies alike to attend to his demands. The tools worked. There were many setbacks and occasional lapses into violence, but the campaign of *satyagraha* succeeded in winning independence for India without any war between the native population and the occupying power. British administrators found Gandhi absurd and exasperating, but they could neither shoot him nor keep him permanently in prison. When he fasted unto death they dared not let him die, knowing that no one who might take his place would be able so well to control the violent temper of his followers. *Satyagraha* was an effective weapon in Gandhi's hands because he was, unlike Tolstoy, an astute politician. For thirty years Gandhi was, in effect, collaborating with the British authorities in keeping India peaceful, while at the same time defying them publicly so that he never appeared to his followers as a British stooge. Successful use of *satyagraha* requires, besides courage and moral grandeur, a talent for practical politics, an understanding of the weak points of the enemy, a sense of humor, and a little luck. Gandhi possessed all these gifts and used them to the full.

Gandhi's luck ran out at the end of his life, when the campaign against British rule was won and he was trying to bring India to independence as a united country. He then had to deal with quarrels between Hindu and Muslim, deeper and more bitter than the power struggle between European and Asian. *Satyagraha* failed to subdue Hindu and Muslim nationalism as it had subdued British imperialism. Five months after the violent birth of independent India and Pakistan, the scene of Jaurès's death was reenacted in Delhi. Like Jaurès, Gandhi was shot by a nationalist who considered him insufficiently patriotic.

With Gandhi, as with Jaurès, died the hope of a continent turning decisively away from war. Nehru, prime minister of newly independent India, had never been a wholehearted believer in nonviolence. The rulers of Pakistan believed in nonviolence even less. Within thirty years after independence, three wars showed how little Gandhi's countrymen had learned from his example. India and Pakistan fought over the disputed province of Kashmir as France and Germany had fought over Alsace and Lorraine. Together with the regiments and warships of the colonial army and navy, the governments of India and Pakistan inherited an addiction to the old European game of power politics. Gandhi's *satyagraha* was an effective weapon for a subject people to use against their oppressors, but his followers discarded it promptly as soon as they gained control of their own government and stood in the oppressors' shoes.

The moral of Gandhi's life and death is that pacifism as a political program is much more difficult to sustain than pacifism as a personal ethic. Being himself a leader of extraordinary charisma and skill, Gandhi was able to organize a whole people around a program of pacifism. He proved that a pacifist resistance movement can be sustained for thirty years and can be strong enough to defeat an empire. The subsequent history of India proved that political pacifism was not strong enough to survive the death of its leader and to withstand the temptations of power.

During the years between the two world wars, while Gandhi was successfully organizing his nonviolent resistance in India, political pacifism was also popular in Europe. European pacifists were encouraged by Gandhi's example and hoped to revive Jaurès's dream of an international alliance of nonviolent resisters against militaristic national governments. The pacifist dream in Europe failed disastrously. There were three main reasons for the failure: lack of leadership, lack of a positive objective, and Hitler. The European pacifists never produced a leader comparable to Gandhi. Einstein was a pacifist, and lent his name and prestige to the pacifist cause until the rise of Hitler led him to change his mind, but he had no wish to be a political leader. Like Tolstoy, he was more of a hero to the world at large than to his own countrymen. Pacifism, even at the peak of Einstein's popularity, was never strong in Germany. It was strongest in England, where George Lansbury, a Christian Socialist with firm pacifist convictions, was leader of the Labour Party from 1931 to 1935. Lansbury was capable of courageous action in the Gandhi style. In 1930, when he was mayor of Poplar in the East End of London, he went to prison rather than submit to government policies which he considered oppressive. He remained a hero to his constituents in East London. But he never attempted to dominate the European scene as Gandhi dominated the scene in India. Gandhi had the tremendous advantage of a positive objective, the cause of Indian independence, around which he could mobilize the enthusiasm of his followers. Lansbury and the other European pacifists had no similar objective; they supported the League of Nations as an international peacekeeping authority, but the League of Nations was an inadequate focus for a mass political movement. The league was widely perceived as nothing more than a debating society for elderly politicians. Nobody could take seriously the picture of millions of Europeans defying their governments in a gesture of loyalty to the league. Gandhi was swimming with the tide of nationalism; Lansbury and his followers were swimming against it. As a result, the foreign policy of the British Labour Party under Lansbury's leadership was wholly negative; no rearmament, no action against Hitler, and no wholehearted commitment to pacifism.

It was Lansbury's fate to preside over the British pacifist movement at the peak of its popularity during the same years which saw Hitler's rise to power in Germany. A few weeks before Hitler became chancellor, the undergraduates of Oxford debated the proposition "That this House will under no circumstances fight for its King and Country," and approved it by a substantial majority. This vote received widespread publicity and may in fact, as the opponents of pacifism later claimed, have encouraged Hitler to pursue his plans of European conquest more boldly. Whether or not Hitler paid attention to the Oxford students' vote, there is no doubt that his aggressive policies were encouraged by the existence of strong pacifist sentiments in England and France. In October 1933, Hitler felt confident enough to withdraw from the international Disarmament Conference which had been meeting before he became chancellor; this action was an official notification to the world that he intended to rearm Germany. Four days later, Lansbury spoke for the Labour Party in the House of Commons:

> We will not support an increase in armaments, but we shall also refuse to support our own or any other government in an endeavour to apply penalties or sanctions against Germany. No one will ask for these if the great nations immediately, substantially disarm and continue until universal disarmament is accomplished.

The great nations were not about to disarm, as Lansbury well knew. His policy meant that England would simply do nothing, neither arm nor disarm. He was caught in the tragic dilemma of political pacifism. The pacifists of England and France, by announcing their unwillingness to fight, made Hitler more reckless in risking war and made the war more terrible when it came. There is no easy answer to this dilemma. A country facing an aggressive enemy must decide either to be prepared to fight effectively or to follow the path of nonviolence to the end. In either case, the decision must be wholehearted and the consequences must be accepted. The example of England in the 1930s proves only that a halfhearted commitment to pacifism is worse than none at all. Halfhearted pacifism is in practice indistinguishable from cowardice. European pacifism became finally discredited when World War II began and halfhearted pacifists could not be distinguished from cowards and collaborators. The debacle of European pacifism has at least one clear lesson to teach us: pacifists, if they are to be effective in the modern world, must be as wholehearted and as brave as Gandhi.

In 1935 Lansbury was forced to choose between his pacifist principles and his position as leader of the Labour Party. Being an honest man, he stuck to his principles and handed over the leadership of the party to Clement Attlee, the same Attlee who became prime minister ten years later and made the decision to arm Britain with nuclear weapons. Pacifism as an effective political force in England was dead. But it was still alive in India. Young Englishmen like me, who were against the establishment and against the empire, acclaimed Gandhi as a hero. We greatly preferred the flamboyant Gandhi to the powerless Lansbury and the colorless Attlee. Our conversation was sprinkled with the rhetoric of pacifist doctrine. If only we had a leader like Gandhi, we said, we would fill the jails and bring the warmongers to their senses. We continued to talk in this style, while Hitler filled his concentration camps in Germany and silenced those who opposed his policies. Then in 1940 Hitler attacked and overran France. We were face to face, as Lansbury had been in 1933, with the classic pacifist dilemma. We still believed theoretically in the ethic of nonviolence, but we looked at what was happening in France and decided that nonviolent resistance would not be effective against Hitler. Reluctantly, we concluded that we had better fight for our King and Country after all.

Forty years later, a book, *Lest Innocent Blood Be Shed*, was written by Philip Hallie, telling the story of a French village which chose the path of nonviolent resistance against Hitler.[2] It is a remarkable story. It shows that nonviolence could be effective, even against Hitler. The village of Le Chambon sur Lignon collectively sheltered and saved the lives of many hundreds of Jews through the years when the penalty for this crime was deportation or death. The villagers were led by their Protestant pastor, André Trocmé, who had been for many years a believer in nonviolence and had prepared them mentally and spiritually for this trial of strength. When the Gestapo from time to time raided the village, Trocmé's spies usually gave him enough warning so that the refugees could be hidden in the woods. German authorities arrested and executed various people who were known to be leaders in the village, but the resistance continued unbroken. The only way the Germans could have crushed the resistance was by deporting or killing the entire population. Nearby, in the same part of France, there was a famous regiment of SS troops, the Tartar Legion, trained and experienced in operations of extermination and mass brutality. The Tartar Legion could easily have exterminated Le Chambon. But the village survived. Even Trocmé himself, by a series of lucky accidents, survived.

Trocmé learned many years later how it had happened that the village survived. The fate of the village was decided in a dialogue between two German soldiers, representing the bright and the dark sides of the German soul. On the one side, Colonel Metzger, an appropriate name meaning "butcher" in German, commander of the Tartar Legion, killer of civilians, executed as a war criminal after the liberation of France. On the other side, Major Schmehling, Bavarian Catholic and decent German officer of the old school. Both Metzger

2. *Lest Innocent Blood be Shed: The Story of the Village of Le Chambon and How Goodness Happened There* (Harper and Row, 1979).

and Schmehling were present at the trial of Dr. Le Forestier, a medical doctor in Le Chambon who was arrested and executed as an example to the villagers. "At his trial," said Schmehling when he met Trocmé later, "I heard the words of Dr. Le Forestier, who was a Christian and explained to me very clearly why you were all disobeying our orders in Le Chambon. I believed that your doctor was sincere. I am a good Catholic, you understand, and I can grasp these things.... Well, Colonel Metzger was a hard one, and he kept on insisting that we move in on Le Chambon. But I kept telling him to wait. I told Metzger that this kind of resistance had nothing to do with violence, nothing to do with anything we could destroy with violence. With all my personal and military power I opposed sending his legion into Le Chambon."

That was how it worked. It was a wonderful illustration of the classic concept of nonviolent resistance. You, Dr. Le Forestier, die for your beliefs, apparently uselessly. But your death reaches out and touches your enemies, so that they begin to behave like human beings. Some of your enemies, like Major Schmehling, are converted into friends. And finally even the most hardened and implacable of your enemies, like the SS colonel, are persuaded to stop their killing. It happened like that, once upon a time, in Le Chambon.

What did it take to make the concept of nonviolent resistance effective? It took a whole village of people, standing together with extraordinary courage and extraordinary discipline. Not all of them shared the religious faith of their leader, but all of them shared his moral convictions and risked their lives every day to make their village a place of refuge for the persecuted. They were united in friendship, loyalty, and respect for one another.

Sooner or later, everybody who thinks seriously about the meaning of war in the modern age must face the question whether nonviolence is or is not a practical alternative to the path we are now following. Is nonviolence a possible basis for the foreign policy of a great country like the United States? Or is it only a private escape route available to religious minorities who are protected by a majority willing to fight for their lives? I do not know the answers to these questions. I do not think that anybody knows the answers. The example of Le Chambon shows us that we cannot in good conscience brush such questions aside. Le Chambon shows us what it would take to make the concept of nonviolent resistance into an effective basis for the foreign policy of a country. It would take a whole country of people standing together with extraordinary courage and extraordinary discipline. Can we find such a country in the world as it is today? Perhaps we can, among countries which are small and homogeneous and possess a long tradition of quiet resistance to oppression. But how about the United States? Can we conceive of the population of the United States standing together in brotherhood and self-sacrifice like the villagers of Le Chambon? It is difficult to imagine any circumstances which would make this possible. But history teaches us that many things which were once unimaginable nevertheless came to pass. At the end of every discussion of nonviolence comes the question which Bernard Shaw put at the end of his play *Saint Joan*:

> O God that madest this beautiful earth, when will it be ready to receive thy Saints? How long, O Lord, how long?

Postscript, 2006

Since this chapter was written in 1984, the emphasis in discussions of war and peace has shifted from national conflicts to the so-called "war against terrorism." In my view, the policy of turning the fight against terrorism into a war is practically ineffective as well as morally wrong. The effective tools for fighting terrorism are civilian police forces and civil defense. Granted that the ends of defeating terrorism are morally justified, it does not follow that the use of war as a means is justified. The "war against terrorism" is probably creating new terrorists faster than it eliminates old ones. To be opposed to this particular war, it is not necessary to be a pacifist.

The story of Le Chambon sur Lignon is told in an excellent documentary film, *Weapons of the Spirit*, produced by Pierre Sauvage in 1987, with many villagers who had been participants in the passive resistance speaking on camera. Sauvage was born in the village while his Jewish parents were hidden there.

11

THE RACE IS OVER

A FEW YEARS ago I walked into a room where there were forty-two hydrogen bombs lying around on the floor, not even chained down, each of them ten times more powerful than the bomb that destroyed Hiroshima. This experience was a sharp reminder of the precariousness of the human condition. It encouraged me to think hard about ways to improve the chances of survival of my grandchildren. Nuclear weapons remain, as George Kennan has said, the most serious danger to mankind and the most serious insult to God.

The disappearance of nuclear weapons from our thinking about the future is a historic change for which we must be profoundly grateful. Fifty years ago and for many years thereafter, nuclear weapons dominated the landscape of our fears. The nuclear arms race was the central ethical problem of our age. Discussion of the ethical dilemmas of scientists centered around bombs and long-range missiles. The evil face of science was personified by the nuclear bomb designer. Now, quietly and unexpectedly, the bombs have faded from our view. But they have not ceased to exist. The danger to humanity of huge stockpiles in the hands of unreliable people is as real as ever. Yet the bombs are not mentioned in our vision of the future. How could this have happened?

In the summer of 1995 I took part in a technical study of the future of the United States' nuclear stockpile. The study was done by a group of academic scientists together with a group of professional bomb designers from the weapons laboratories. The purpose of the study was to answer a question: Would it be technically feasible to maintain forever a stockpile of reliable nuclear weapons of existing designs without further nuclear tests? The study did not address the underlying political questions, whether reliable nuclear weapons would always be needed and whether further nuclear tests would always be undesirable. Each of us had private opinions about the political questions, but politics was not the business of our study. We assumed as the ground rule for the study that the weapons in the permanent stockpile must be repaired and remanufactured without change in design as their components deteriorate and decay. We assumed that the new components would differ from the old ones when replacements were made, because the factories making the old components would no longer exist. We looked in detail at each type of weapon and checked that its functioning was sufficiently robust so that minor changes in the components would not cause it to fail. We concluded our study with a unanimous report, saying that a permanently reliable nuclear stockpile without nuclear testing is feasible. Unanimity was essential.

Unanimity was made possible by the objectivity and the personal integrity of the four weapons designers who worked side by side with us for seven weeks, John Richter and John Kammerdiener from Los Alamos, Seymour Sack from Livermore, and Robert Peurifoy from Sandia. They are impressive people, master craftsmen of a demanding technology. They have spent the best part of their lives planning and carrying out bomb tests. They remember every test, whether it succeeded or failed. They know why each test was done, and what was learned from its success or failure. Their presence was essential to our work, and their names on the report gave credibility to our conclusions. They are survivors of a vanishing culture. They lived through the heroic age of weapon-building. They will not and cannot be replaced. By working on this study, they unselfishly helped our country to move safely into a world in which people with the special qualities and talents of these four men will no longer be needed.

The conclusion of our study was a historical landmark, commemorating the fact that the nuclear arms race is finally over. The nuclear arms race raged with full fury for only twenty years, the 1940s and 1950s. Then it petered out slowly for the next thirty years, in three stages. The science race petered out in the 1960s, after the development of highly efficient hydrogen bombs. Nuclear weapons then ceased to be a scientific challenge. The military race petered out in the 1970s, after the development of reliable and invulnerable missiles and submarines. Nuclear weapons then ceased to give a military advantage to their owners in real-world conflicts. The political race petered out in the 1980s, after it became clear to all concerned that huge nuclear weapons industries were environmentally and economically disastrous. The size of the nuclear stockpile then ceased to be a political status symbol. Arms control treaties were concluded at each stage, to ratify with legal solemnity the gradual petering out of the race. The atmospheric test ban of 1963 ratified the end of the science race, the ABM and SALT treaties of the 1970s ratified the end of the military race, and the START treaties of the 1980s ratified the end of the political race.

How may we extrapolate from this history into the world of the 1990s and beyond? The security and the military strength of the United States now depend primarily on nonnuclear forces. Nuclear weapons are on balance a liability rather than an asset. The security of the United States will be enhanced if all deployments of nuclear weapons, including our own, are gradually reduced to zero. For the next fifty years we should attempt to drive the nuclear arms race in reverse gear, to persuade our allies and our enemies that nuclear weapons are more trouble than they are worth. The most effective moves in this direction are unilateral withdrawals of weapons. The move that signaled the historic shift of the arms race into reverse gear was the unilateral withdrawal of land-based and sea-based tactical nuclear weapons by President George Bush in 1991. Chairman Mikhail Gorbachev responded quickly with similarly extensive withdrawals of Soviet weapons. The testing moratorium of 1992 was another effective move in the same direction.

To drive the nuclear arms race further in reverse gear, we need to pursue three long-range objectives: worldwide withdrawal and destruction of weapons, complete cessation of nuclear testing, and an open world in which nuclear activities of all countries are to some extent transparent. In pursuing these objectives, unilateral moves are usually more persuasive than treaties. Unilateral moves tend to create trust, whereas negotiation of treaties often tends to create suspicion.

Our nuclear stockpile study fitted well into the context of the reverse-gear arms race. The purpose of the study was to achieve a technical stabilization of our stockpile, to clarify what needs to be done to maintain a limited variety of weapons indefinitely without testing. Stabilization is the essential prerequisite for allowing the weapons to disappear gracefully. Once a stable regime of stockpile maintenance has been established, the weapons will attract less attention both nationally and internationally. They will acquire the qualities that a stable nuclear deterrent force should have: awesomeness, remoteness, silence. Gradually, as the decades of the twenty-first century roll by, these weapons will become less and less relevant to the problems of international order in a hungry and turbulent world. The time may come when nuclear weapons are perceived as useless relics of a vanished era, like the horses of an aristocratic cavalry regiment, maintained only for ceremonial purposes. When nuclear weapons are generally regarded as absurd and irrelevant, the time may have come when it will be possible to get rid of them altogether.

The time when we can say good-bye to nuclear weapons is still far distant, too far to be clearly envisaged, perhaps a hundred years away. Until that time comes, we must live with our weapons as responsibly and as quietly as we can. That was the purpose of the stockpile study, to make sure that our weapons can be maintained with a maximum of professional competence and a minimum of fuss and excitement, until in the fullness of time they will no longer be considered necessary. In the meantime, the ethical dilemmas concerned with nonnuclear weapons and nonnuclear warfare remain unresolved.

The abolition of war is an ultimate goal, more remote than the abolition of nuclear weapons. The idea espoused early in the nuclear age by J. Robert Oppenheimer, that the existence of nuclear weapons might lead to the abolition of war, turned out to be an illusion. The abolition of war is a prime example of an ethical problem that science is powerless to deal with. The weapons of nonnuclear war, guns and tanks and ships and airplanes, are available on the open market to anybody with money to pay for them. Science cannot cause these weapons to disappear. The most useful contribution that science can make to the abolition of war has nothing to do with technology. The international community of scientists may help to abolish war by setting an example to the world of practical cooperation extending across barriers of nationality, language, and culture.

Postscript, 2006

The Stockpile Stewardship Program has survived the political upheavals of the nine years since this piece was written. The policy of replacing old weapons without changing their design has been maintained. But some influential people are now advocating a change of policy, to replace old weapons with a Reliable Replacement Warhead (RRW). The RRW has a new design, making it simpler, more rugged, and less affected by aging. The new policy would make technical sense, but it would be politically disastrous, encouraging other countries to introduce new types of warheads, and going against the long-range goal of letting nuclear weapons fade out gracefully.

12

THE FORCE OF REASON

JOSEPH ROTBLAT HAS devoted the greater part of his long life to the struggle to eliminate nuclear weapons from the earth. Unfortunately he was still in Poland in January 1939 when the possibility of nuclear weapons first became generally known. He was aware of the possibilities, but his voice was not heard in the public discussions of that year. If his voice had been heard, it is possible that history might have taken a different course. In 1939 a great opportunity was missed. That year was the last chance for physicists to establish an ethical tradition against nuclear weapons, similar to the Hippocratic tradition that stopped biologists from promoting biological weapons. The chance was missed, and from that point on the march of history led inexorably to Hiroshima.

In January 1939 a meeting of physicists was held at George Washington University in Washington, D. C. The meeting had been planned by George Gamow long before fission was discovered. It was one of a regular series of annual meetings. It happened by chance that Niels Bohr arrived in America two weeks before the meeting, bringing from Europe the news of the discovery of fission. Gamow quickly reorganized the meeting so that fission became the main subject. Bohr and Enrico Fermi were the main speakers. For the first time, the splitting of the atom was publicly described, and the consequent possibility of atomic bombs was widely reported in newspapers. Not much was said at the meeting about atomic bombs. Everyone at the meeting was aware of the possibilities, but nobody spoke up boldly to suggest that questions of ethical responsibility be put on the agenda. The meeting came too soon for any consensus concerning ethical responsibilities to be reached. Most of the people at the meeting were hearing about fission for the first time. But it would have been possible to start a preliminary discussion, to make plans for an informal organization of physicists, and to prepare for further meetings. After several weeks of preparation, a second meeting might have been arranged with the explicit purpose of reaching an ethical consensus.

Within a few months after the January meeting, Bohr and John Wheeler had worked out the theory of fission in America, the possibility of a fission chain reaction had been confirmed by experimenters in several countries, by Rotblat in Poland among others, and Yakov Zeldovich and Yuli Khariton had worked out the theory of chain reactions in Russia. All this work was openly discussed and rapidly published. The summer of 1939 was the moment for decisive action to forestall the building of nuclear weapons. Nothing was then officially secret. The leading actors in all countries, Bohr and Einstein and Fermi and Werner Heisenberg and Pyotr Kapitsa and Khariton and Igor Kurchatov and Frédéric Joliot and Rudolf Peierls and J. Robert Oppenheimer, were still free to talk to one another and to decide upon a common course of action. The initiative for such a common course of action would have most naturally come from Bohr and Einstein. They were the two giants who had the moral authority to speak for the conscience of mankind. Both of them were international figures who stood above narrow national loyalties. Both of them were not only great scientists but also political activists, frequently engaged with political and social problems. Why did they not act? Why did they not at least try to achieve a consensus of physicists against nuclear weapons before it was too late? Perhaps they would have acted, if Joseph Rotblat had been there to urge them on.

Thirty-six years later, the sudden discovery of recombinant DNA technology presented a challenge to biologists, similar to the challenge which the discovery of fission had presented to physicists. The biologists promptly organized an international meeting at Asilomar, at which they hammered out an agreement to limit and regulate the uses of the dangerous new technology. It took only a few brave spirits, with Maxine Singer in the lead, to formulate a set of ethical guidelines which the international community of biologists accepted. What happened at George Washington University in 1939 was quite different. No brave spirits emerged from the community of physicists at the meeting. Instead of coming together to confront the common danger facing humanity, the two leading figures, Bohr and Fermi, began to argue about scientific credit. Fermi read aloud a telegram that he received during the meeting from his colleague Herb Anderson at Columbia University, announcing the successful verification of the fission process by direct detection of the pulses of ionization produced by fission fragments. Bohr objected to the claim of credit for Anderson, and pointed out that the same experiment had been done earlier by Otto Frisch in his own institute in Copenhagen. Bohr was worried because Frisch's letter to *Nature* reporting his experiment had not yet appeared in print. Fermi was fighting for his friend Anderson and Bohr was fighting for his friend Frisch. Scientific priority was more important than common danger. The habit of fighting for priority, as prevalent in the scientific community of the 1930s as it is today, was hard to break. Neither Bohr nor Fermi was able to rise above these parochial concerns. Neither of them felt any urgent need to deal with the larger issues that fission had raised.

As soon as Hitler overran Poland in September 1939 and World War II began, the chance of achieving a tacit agreement of physicists in all countries not to build nuclear weapons disappeared. We know why the physicists in Britain and America felt compelled to build weapons. They were afraid of Hitler. They knew that fission had been discovered in Germany in 1938 and that the German government had started a secret uranium project soon thereafter. They had reason to believe that Heisenberg and other first-rate German scientists were involved in the secret project. They had great respect for Heisenberg, and equally great distrust. They were desperately afraid that the Germans, having started their project earlier, would succeed in building nuclear weapons first. They believed that America and Britain were engaged in a race with Germany which they could not afford to lose. They believed that if Hitler got nuclear weapons first he could use them to conquer the world. Joseph Rotblat was marooned in Britain with his homeland destroyed and his wife in mortal danger. He had more reason than anybody else to be afraid of what Hitler might do with nuclear weapons.

The fear of Hitler was so pervasive that hardly a single physicist who was aware of the possibility of nuclear weapons could resist it. The fear allowed scientists to design bombs with a clear conscience. In 1941 they persuaded the British and American governments to build the factories and laboratories where bombs could be manufactured. It would have been impossible for the community of British and American physicists to say to the world in 1941, "Let Hitler have his nuclear bombs and do his worst with them. We refuse on ethical grounds to have anything to do with such weapons. It will be better for us in the long run to defeat him without using such weapons, even if it takes a little longer and costs us more lives." Hardly anybody in 1941 would have wished to make such a statement. Even Rotblat would not have made such a statement. And if some of the scientists had wished to make it, the statement could not have been made publicly, because all discussion of nuclear matters was hidden behind walls of secrecy. The world in 1941 was divided into armed camps with no possibility of communication between them. Scientists in Britain and America, scientists in Germany, and scientists in the Soviet Union were living in separate black boxes. It was too late in

1941 for the scientists of the world to take a united ethical stand against nuclear weapons. The latest time that such a stand could have been taken was in 1939, when the world was still at peace and secrecy not yet imposed.

With the benefit of hindsight, we can now see that if the physicists in 1939 had quietly agreed not to push the development of nuclear weapons in their various countries, there was a good chance that the weapons would not have been developed anywhere. In every country it was the scientists and not the political leaders who took the initiative to begin nuclear weapons programs. Hitler, as we learned afterward, was never seriously interested in nuclear weapons. The Japanese military leaders were not seriously interested. Stalin was not seriously interested until he was secretly informed of the size and seriousness of the American program. Roosevelt and Churchill only became interested after their scientific advisers pushed them into it. If the scientific advisers had refrained from pushing, it is likely that World War II would have ended without any Manhattan Project and without any Soviet equivalent. It would then have been possible, as soon as the war was over, to begin negotiations among the victorious allies to establish a nuclear-weapon-free world with some hope of success. We cannot know whether this road not taken would have avoided the nuclear arms race altogether. At least it would have been a saner and wiser road than the one we followed.

In October of 1995, I was giving a lunchtime lecture to a crowd of students at George Washington University about the history of nuclear weapons. I told them about the meeting that had been held in a nearby building on their campus in January 1939. I told them how the scientists at the meeting missed the opportunity that was fleetingly placed in their hands, to forestall the development of nuclear weapons and to change the course of history. I talked about the nuclear projects that grew during World War II, massive and in deadly earnest in America, small and halfhearted in Germany, serious but late-starting in Russia. I described the atmosphere of furious effort and intense camaraderie that existed in wartime Los Alamos, with the British and American scientists so deeply engaged in the race to produce a bomb that they did not think of stopping when the opposing German team dropped out of the race. I told how, when it became clear in 1944 that there would be no German bomb, only one man, of all the scientists in Los Alamos, stopped. That man was Joseph Rotblat. I told how Rotblat left Los Alamos and became the leader of the Pugwash movement, working indefatigably to unite scientists of all countries in efforts to undo the evils to which Los Alamos gave rise. I remarked how shameful it was that the Nobel Peace Prize, which had been awarded to so many less deserving people, had never been awarded to Rotblat. At that moment one of the students in the audience shouted, "Didn't you hear? He won it this morning." I shouted, "Hooray," and the whole auditorium erupted in wild cheering. In my head the cheers of the students are still resounding.

Postscript, 2006

Joseph Rotblat died in 2005 at the age of ninety-six. He shared the 1995 Nobel Peace Prize with the Pugwash organization, which he had founded and served as director-general for many years. He continued to be vigorously active until a few weeks before his death.

13

THE BITTER END

ARMAGEDDON[1] IS A mosaic composed of hundreds of brightly colored fragments, each one a story told by an eyewitness. Most of the fragments occupy less than a page. The mosaic is a panorama of the last eight months of World War II in Europe, between September 1944 and May 1945. These were the months in which British and American armies in the West and Russian armies in the East fought their way across the frontiers of Germany and finally defeated German armies on German soil. The panorama is remarkable in many ways. The toll of death and destruction and misery during these eight months was unequaled by any similar period in the long history of human misfortunes, wars, and persecutions. The German armies fought with extraordinary skill and bravery to defend their shrinking territory, long after any realistic hope of victory had disappeared. The invading armies, in spite of profound political and cultural differences, succeeded in working together until their job was done. Each of these aspects of the panorama is illuminated by personal experiences described in the individual fragments.

The eyewitnesses are divided more or less evenly between soldiers and civilians, between males and females, between Germans, Russians,

---

1. Max Hastings, *Armageddon: The Battle for Germany, 1944–1945* (Knopf, 2004).

Poles, Jews, Britons, and Americans. Max Hastings interviewed most of them personally during the year 2002, when he traveled to their countries and met them in their homes, most of them by then old people recalling events that happened when they were in their teens or twenties. Hastings is well aware that memories recalled after fifty-eight years are unreliable. As he says, these memories are not history. They are the raw material out of which history may grow. They provide a useful corrective to official histories based on written documents, which may be equally unreliable. They give us direct access to the human face of war, the face that the official histories usually ignore.

To interview German and Russian witnesses, Hastings used interpreters whose help he gratefully acknowledges. The interpreters not only translated but also helped him to find witnesses with good stories to tell, and these witnesses then led him to others among their friends and acquaintances. Two groups of witnesses that he found in this way were Russian women who had been girl-soldiers in the Red Army, and German women who had been refugees escaping from East Prussia when the Red Army overran their homeland. The Russians describe a tough but in many ways joyful atmosphere of comradeship and shared hardship on the road to victory. The Germans describe a nightmare of death and destruction as they made their way as exiles from a lost paradise. It is not surprising that the best witnesses are usually female, since women live longer than men in all countries, and especially in Russia.

In addition to the recent interviews, Hastings also includes in his account interviews that he recorded long ago as raw material for his other historical books, *Bomber Command*, a history of the British strategic bombing of Germany published in 1979, and *Overlord*, a history of the invasion of France by British and American armies published in 1984. The earlier interviews are mostly with senior commanders and politicians who were no longer alive when *Armageddon* was written. Hastings also includes quotations from letters and documents that he found in Russian archives and in various other archives that recently became accessible to historians.

The older interviews and letters provide a striking contrast to the newer interviews. The older sources show us war as seen by commanders and planners, a succession of operations following one another in a logical sequence like the moves in a game of strategy. The new interviews show us war as seen by foot soldiers and civilian victims, a succession of murderous assaults that occur randomly and unpredictably, without any intelligible pattern. Both views of war are valid, and both are necessary components of any history that attempts to be truthful. Hastings keeps the two views in balance and blends them skillfully as he builds his mosaic. Where the two views conflict, he tends to give greater credence to the foot soldier than to the general.

My own limited experience of World War II leads me to share Hastings's bias in favor of foot soldiers. I belong to the same generation as Hastings's foot-soldier witnesses. I was lucky not to be a foot soldier. I was a civilian living in London at various times when German bombers were flying overhead. From time to time a bomb would fall and demolish a couple of houses. Our antiaircraft guns made a lot of noise but I never saw them hit an airplane. I remember thinking that the German kids overhead were probably as bewildered as I was. The nearest I came to being hurt was in January 1944, when a bomb fell on our street and broke our windows. This happened while the German army in Russia was fighting monstrous battles to hold its ground against the Soviet winter offensive. The fate of the world was being decided in Russia.

Hitler was evidently out of touch with reality, sending his precious airplanes to London to break our windows instead of sending them to Russia where they were desperately needed. The most vivid impression that remains to me from those times is a feeling of irrelevance. The little game that I was witnessing in London was wholly irrelevant to the serious war that we were supposed to be fighting. My memory fits well with the picture of the war that Hastings shows us. The serious and purposeful fighting is done by a small fraction of the people involved. Most of the people, most of the time, are irrelevant. Irrelevant or not, they still suffer the consequences.

The history of World War II teaches us several lessons that are still valid today. First is the immense importance of the Geneva conventions on humane treatment of prisoners in mitigating the human costs of war. All through Hastings's narrative, we see a stark contrast between two kinds of war, the war in the West following the Geneva rules and the war in the East fought without rules. A large number of witnesses of the western war, German as well as British and American, owe their lives to the Geneva conventions. In the western war, soldiers fought hard as long as fighting made sense, and surrendered when fighting did not make sense, with a good chance of being treated decently as prisoners of war. Many of the prisoners on both sides were killed in the heat of battle before reaching prison camps, but most of them survived. Those who reached the prison camps were treated in a civilized fashion, with some supervision by delegates of the International Red Cross. They were neither starved nor tortured.

At the same time, on the eastern side of the war, brutality was the rule and the International Red Cross had no voice. Civilians were routinely raped and murdered, and prisoners of war were starved. Soldiers were expected to fight to the death, and most of them did, since they had little hope of survival as prisoners. It is not possible to calculate the numbers of lives saved in the West and lost in the East by following and not following the Geneva rules. The numbers certainly amount to hundreds of thousands in the West and millions in the East. Americans who are trying today to weaken or evade the Geneva rules are acting shortsightedly as well as immorally.

A second important lesson of World War II is the fact that German soldiers consistently fought better than Britons or Americans. Whenever they were fighting against equal numbers, the Germans always won, a fact recognized by the Allied generals, who always planned to achieve numerical superiority before attacking. This was the main reason why the Allied advance into Germany was slow. If the Allied soldiers had been able to fight like Germans, the war would probably have been over in 1944 and millions of lives would have been saved.

Hastings explains the superiority of German soldiers as a consequence of the difference between a professional army and a citizen army. The Germans were professionals, brought up in a society that glorified soldiering, and toughened by years of fighting in Russia. The British and American soldiers were mostly amateurs, civilians who happened to be in uniform, brought up in societies that glorified freedom and material comfort, and lacking experience of warfare. The difference between the German and Allied armies was similar to the difference between Southern and Northern armies in the American Civil War. The Southern soldiers fought better and the Southern generals were more brilliant. The Northern soldiers won in the end because there were more of them and they had greater industrial resources, just as the Allies did in World War II. The leaders of the Old South romanticized war and led their society to destruction, just as the leaders of Germany did eighty years later.

Hastings says we should take pride in the fact that our soldiers did not fight as well as Germans. To fight like Germans, they would have had to think like Germans, glorifying war and following their leaders blindly. We should consider ourselves lucky that soldiering is not embedded in our culture as it was embedded in the culture of Germany in 1944. The Germans who survived World War II are also lucky, since the devastation of their country finally convinced them that soldiering was a false god.

The third lesson of World War II is the value of international alliances. International alliances are slow and cumbersome and unromantic. Leaders of international alliances cannot move quickly. They must make compromises and accept delays in order to achieve consensus. They cannot make brilliant and disastrous decisions as Hitler did. They cannot lead their people to destruction. To fight a war within the constraints of an international alliance is a good protection against fatal mistakes and follies. Eisenhower was an ideal person to lead an international alliance. He was a mediocre strategist and an excellent diplomat. He had no interest in military glory. His priorities were to hold the alliance together and to win the war with the minimum number of casualties. Unlike the brilliant German generals who were his opponents, he demanded as little as possible from his soldiers. He preferred to end the war with live soldiers rather than with dead heroes.

Eisenhower won the war by going slow and avoiding big mistakes. The most important decision that he made during the period covered by *Armageddon* was to send a personal message informing Stalin that his armies would not try to take Berlin. The message was sent in March 1945, without consulting the political authorities in Washington and London. Eisenhower knew that several of his subordinate generals wanted passionately to march in triumph through Berlin. He knew that the attempt to do so might result either in a bloody battle with the Germans or in a disastrous clash with the Russians. He knew that many political leaders in Washington and London would give strong support to a grab for Berlin. He took personal responsibility for a decision that would be politically unpopular at home but would save the alliance with Russia and incidentally save the lives of his soldiers.

Hastings in his penultimate chapter, "The Earth Will Shake as We Leave the Scene," describes how the war in the East ended. The title of the chapter is a quote from Joseph Goebbels, spoken shortly before he committed suicide. Stalin launched his final offensive against Berlin in April 1945 and lost 350,000 men in three weeks. The Germans lost about a third as many before they were overrun. The British and Americans stopped at the Elbe River and came home alive.

I remember a conversation with my father in 1940, when France had dropped out of World War II and England was fighting alone against Germany. I was depressed and despondent, but my father was disgustingly cheerful. I said the situation was hopeless, there was no way we could win the war, and we had only the choice between surrendering and continuing to fight forever. My father said, don't worry, just hang on, and things will turn out all right in the end. He said, all we have to do is to behave halfway decently, and the whole world will come to our side. I did not believe him, but of course he was right. We did behave halfway decently, and within two years the whole world came to our side. Instead of carrying the fate of the world on our shoulders, we became minor players in a grand alliance. The alliance took away our freedom of action, but allowed us to achieve our objectives at a reasonable cost.

The war that is now raging in Iraq illustrates once again the value of international alliances. If the decision to go to war had been in the hands of an international alliance, the war would probably never have started. If it had started by deliberate decision of an international authority, it would have been a war of limited objectives like the first Gulf War of 1991. It would have left a functioning government in Baghdad responsible for maintaining peace and security. The United States would have avoided the disastrous mistakes that are always more likely to occur when actions are taken hastily and unilaterally.

A fourth lesson of World War II is the moral ambiguity of war even when it is fought for a good cause. *Armageddon* is full of examples of moral ambiguity, both at the level of individual soldiers and at the level of governments. No matter whether their cause is just or unjust, individual soldiers in the heat of battle frequently kill prisoners of war or innocent bystanders. Women are raped, goods are stolen, and homes are destroyed. Horror stories are more horrible in the East but also occur in the West. Those who commit crimes are not always German. War is inherently immoral, and everyone who engages in war is doing things which under normal circumstances would be considered criminal. One of Hastings's witnesses was a private in an American infantry division during the German offensive in the Ardennes in December 1944. Speaking of German prisoners, he says, "If they wore the black uniforms of the SS, they were shot." He did not know that all German tank crews had black uniforms, whether they belonged to SS or to regular army units.

At the level of governments, there are two egregious examples of moral ambiguity, the betrayal of Poland and the strategic bombing of German cities. Poland was a moral problem for the Allies from the beginning of the war to the end. At the beginning, Britain and France declared war on Germany when Hitler invaded Poland, but took no military action in the West while Poland was overrun. Stalin had signed an agreement with Hitler to divide Poland between Germany and Russia. Britain and France were legally and morally obliged to defend Poland, but gave the Poles no help. During the years between 1941 and 1944, when Poland was occupied by Germany, airplanes with Polish crews were flying from bases in Britain to drop supplies and weapons to resistance fighters in Poland. These "special operations" to Poland suffered terrible losses, averaging 12 percent per operation. They were suicide missions for the crews that flew them. They provided minimal help to the resistance.

When the resistance fighters rose in revolt against the Germans in Warsaw in August 1944, the Allies again did nothing to help, and the Germans crushed the revolt mercilessly. Hastings found few witnesses of the catastrophe in Warsaw, since hardly any of the resistance fighters survived. Soon after that, Soviet troops occupied Poland and installed their own puppet government, with enforcement provided by the Soviet secret police. The final act of betrayal was the Yalta agreement of February 1945, in which Roosevelt and Churchill agreed, in effect, to let Stalin do what he wished with Poland. Britain and America were faced with an insoluble moral dilemma. To defeat Hitler, they needed to maintain the alliance with Stalin. To maintain the alliance, they needed to abandon Poland.

The moral issues raised by the strategic bombing of German cities are less clear-cut. The main question is whether the bombing of cities was morally justified as a military operation helping to win the war. Hastings devotes a long chapter, "Firestorms: War in the Sky," to the bombing campaign, with testimony from many witnesses who were flying in the bombers and others who were among the bombed. He lets the witnesses speak for themselves. They do not have much to say about the moral issues. The bomber crewmen still believe what they were told by their commanders, that the bombing was morally justified since it made a major contribution to winning the war. German civilian witnesses still mostly consider themselves victims of an evil and misdirected vengeance. Prisoners and slave laborers in Germany welcomed the bombing as a promise of their approaching liberation. Since I was myself a witness, serving as a civilian analyst at the headquarters of the Royal Air Force Bomber Command from which the British part of the campaign was directed, I add my testimony here to the others.

At Bomber Command headquarters, I was responsible for collecting and analyzing information about bomber losses. Our losses were tremendous, more than 40,000 highly trained airmen killed. Until the last few months of the war, a crewman had only one chance in four of surviving to the end of his tour of thirty operations. Many of the survivors signed on for a second tour, in which their chances of survival were not much better. The total economic cost of Bomber Command, including the production of airplanes and fuel and bombs, the training of crews, and the conduct of operations, was about one quarter of the entire British war effort. It was my judgment at the time, and remains so today, that the cost of Bomber Command in men and resources was far greater than its military effectiveness. From a military standpoint, we were hurting ourselves more than we were hurting the Germans. It cost us far more to attack German cities than it cost the Germans to defend them. The German night-fighter force, which was the most effective part of the defense and caused most of our losses, was minuscule compared with Bomber Command.

There is overwhelming evidence that the bombing of cities strengthened rather than weakened the determination of the Germans to fight the war to the bitter end. The notion that bombing would cause a breakdown of civilian morale turned out to be a fantasy. And the notion that bombing would cause a breakdown of weapons production was also a fantasy. After a devastating attack on a factory, the Germans were able to repair the machinery and resume full production in an average time of six weeks. We could not hope to attack the important factories frequently enough to keep them out of action. We learned after the war that, in spite of the bombing, German weapons production increased steadily up to September 1944. In the last few months of the war, bombing of oil refineries caused the German armies to run out of oil, but they never ran out of weapons. Putting together what I saw at Bomber Command with the testimony of Hastings's witnesses, I conclude that the contribution of the bombing of cities to military victory was too small to provide any moral justification for the bombing.

Unfortunately, the official statements of the British government always claimed that the bombing was militarily effective and therefore morally justified. As a result of their ideological commitment to bombing as a war-winning strategy, the leaders of the government were deluding themselves and also deluding the British public. Hastings says that in the last phase of the war "the moral cost of killing German civilians in unprecedented numbers outweighed any possible strategic advantage." I would make a stronger statement. I would say that quite apart from moral considerations, the military cost of killing German civilians outweighed any possible strategic advantage.

The strategic thinking of all the participants in World War II was dominated by their experiences in World War I. Memories of World War I were handed down from the parents to the children of that generation. Paradoxically, the winners and losers of World War I derived opposite conclusions from their experiences. The winners, Britain and America and France, looked back on World War I as an unmitigated horror. Their strategies in World War II were driven by the imperative that the horrors of World War I must not be repeated. For Britain and America, the key to victory was to be strategic bombing. For France, the key was a defensive strategy based on the Maginot Line. But the losers, Germany and Russia, looked back on World War I as a heroic struggle which they could have won if they had had more competent and resolute political leadership. Their strategies in World War II were driven by the idea that they could fight World War I over again and this time do it right. The key to victory was a great army organized to carry out enormous offensive operations, like the German offensive that almost overran Paris and the Russian offensive that almost overran East Prussia in 1914, but this time with better training and better equipment so that there would be no "almost." This strategy succeeded for the Germans in France in 1940 and for the Russians in East Prussia in 1945.

Hastings's book describes how these different strategies partially succeeded and partially came to grief in the bloody finale of World War II. While the British and American armies were cautiously moving into Germany, the Germans and Russians were fighting World War I over again, launching large-scale offensives and counteroffensives, accepting huge losses on both sides, as the Red Army fought its way from the Vistula to the Elbe. Two huge Russian armies raced one another to be the first to march in triumph through Berlin. The price that this race cost in dead and wounded was willingly paid. Meanwhile, the Americans and British failed to defeat Germany with bombing, but succeeded in avoiding the catastrophic carnage of World War I.

One of the notorious examples of the tragic waste of human life in World War I was the death of Henry Moseley, a brilliant young physicist who made a great discovery in 1913 and then died as a volunteer soldier at Gallipoli in 1915. The British government made a deliberate decision in World War II not to allow scientific talent to be wasted. As a result of this decision, I was given a safe job as a statistician at Bomber Command, while my contemporaries who flew in the bombers mostly died. I owe my survival directly to Henry Moseley and to the British strategy of minimizing the losses of scientists. If the authorities had not clung so stubbornly to their belief in the effectiveness of strategic bombing, they could have saved not only me but the others too.

After *Armageddon* was written, another book by a witness of the German tragedy was published in English, *The End*[2] by Hans Nossack. Nossack was a famous German writer who was living in Hamburg during World War II. The city was destroyed in July 1943 by massive incendiary attacks, culminating in a firestorm similar to the one that destroyed Dresden in 1945. The destruction of Hamburg was the most successful of all the operations of the British Bomber Command. Nossack was taking a holiday in a village near Hamburg when it happened. After the firestorm, Nossack walked through the city and recorded what he saw. His book was written in November 1943 and published in German as part of a longer work with the title *Interview mit dem Tode* (*Interview with Death*) in 1948. The English version is elegantly translated by Joel Agee, and illustrated with photographs taken after the catastrophe in 1943 by Erich Andres. Agee has added a foreword describing the history of the book and the translation. The book is a work of art, distilling into sixty-three short pages the German experience of total destruction, just as John Hersey's *Hiroshima* distilled the Japanese experience three years later. It is unfortunate that the publication of Agee's translation was delayed by thirty years.

---

2. Hans Erich Nossack, *The End: Hamburg 1943*, translated from the German and with a foreword by Joel Agee, and with photographs by Erich Andres (University of Chicago Press, 2004).

The End was written only four months after the events that it describes, before the Allied invasion of France and long before the end of the war. It gives authentic testimony, untainted by knowledge of later events, of the effect of strategic bombing on a civilian population. It describes briefly the physical horrors of the cleanup after the bombing:

> People said that the corpses, or whatever one wants to call the remains of dead people, were burned on the spot or destroyed in the cellars with flamethrowers. But actually, it was worse. The flies were so thick that the men couldn't get into the cellars, they kept slipping on maggots the size of fingers, and the flames had to clear the way for them to reach those who had perished in flames.
>
> Rats and flies were the lords of the city. Insolent and fat, the rats disported themselves on the streets.

But Nossack was not so much concerned with physical horrors as with the state of mind of the survivors. According to his testimony, the survivors mostly returned to live in the cellars of their ruined homes and started as soon as possible to resume their accustomed routines. They preferred to live in caves among friends rather than in houses among strangers. The struggle to survive kept them busy and gave them little time for grieving. Since they had lost everything, all they had left was each other. They shared what little they had, and worked together to bring the city back to life.

Concerning the question whether the bombing increased or decreased the loyalty of citizens to the government, Nossack has this to say:

> It would be a mistake, however, to speak of latent unrest and rebellion at the time. Not only the enemies but also our own authorities miscalculated in this respect. Everything went on

> very quietly and with a definite concern for order, and the State took its bearings from this order that had arisen out of the circumstances. Wherever the State sought to impose regulations of its own, people just got upset and angry.... Today the State credits itself with having exercised "restraint," but that is ridiculous. Others say we were much too apathetic at the time to be capable of revolt. That is not true either. In those days everyone said what was on his mind, and no feeling was further from people than fear.

Nossack's conclusion is that the bombing decreased the respect of citizens for the State but increased their loyalty to the community.

Concerning the question whether the bombing was criminal, Nossack says:

> I have not heard a single person curse the enemies or blame them for the destruction. When the newspapers published epithets like "pirates of the air" and "criminal arsonists," we had no ears for that. A much deeper insight forbade us to think of an enemy who was supposed to have caused all this; for us, he, too, was at most an instrument of unknowable forces that sought to annihilate us. I have not met even a single person who comforted himself with the thought of revenge. On the contrary, what was commonly said or thought was: Why should the others be destroyed as well?

Nossack expresses his own astonishment that people accepted their fate with stoic spirit, as if the destruction were not the work of human hands but of an impersonal destiny.

The End gives us an intimate picture of Armageddon as it was experienced by an individual German. The German tradition in life and literature is intensely philosophical. More than other people,

Germans isolate themselves from reality by spinning cocoons of philosophy around unpleasant facts. Nossack describes himself walking through the ruins of Hamburg like a disembodied spirit, detached from the things and people that he is observing. He writes:

> We walked through the world like dead men who no longer care about the petty miseries of the living.... If after hours of searching you met a person, it would only be someone else wandering in a dream through the eternal wasteland. We would pass each other with a shy look and speak even more softly than before.

Perhaps this habit of philosophical detachment helps to explain why the German armies fought so professionally to the bitter end in 1945, when every day that they prolonged the fighting only increased the suffering of their own people as well as the suffering of the others.

Postscript, 2006

This review provoked an unusually heavy volume of letters in response, some approving and some disapproving. I am indebted to Martin Gaynes for correcting the most serious error in my account. I wrote that in the western theater of war, those prisoners who reached the prison camps were treated in a civilized fashion.

This was no longer true in the final months of the war. To set the record straight, here is an extract from Martin Gaynes's letter:

> In December 1944, thousands of American soldiers captured during the Battle of the Bulge were transported to Stalag IX-B, the largest German prisoner-of-war camp, near Frankfurt, Germany. A military order was issued that all Jewish soldiers identify themselves. After the Americans refused to comply, Nazi

> guards selected the GIs they thought looked Jewish, had Jewish-sounding last names, or whom they classified as undesirables. Less than a third of the American soldiers selected were in fact Jewish. Packed into railway boxcars with no food, water, or toilets, they were transported further into the German countryside. Five days later they arrived at Berga, a satellite of the concentration camp at Buchenwald. The Americans were put to work alongside European concentration camp prisoners.... Many died of injuries, malnutrition, disease, and exhaustion. Several were fatally shot by guards for no apparent reason. Some went insane. By April 1945, as the Allies advanced, the SS ordered the evacuation of the camp. Surviving prisoners were marched through rain, snow and bitter cold on a 150-mile procession of death.... The nightmare finally ended on April 23, 1945, when advancing American units came upon and liberated the final surviving prisoners.

As I remarked in my letter thanking Martin Gaynes for his correction of my mistake: "The fact that there was a major breach of the Geneva rules in the treatment of Jewish prisoners does not lessen the value of the rules for saving lives, either in World War II or today."

III

History of Science and Scientists

14

TWO KINDS OF HISTORY

WE ARE DOUBLY lucky. A thoughtful and sensitive book, Yuri Manin's *Mathematics and Physics*, has been thoughtfully and sensitively translated.[1] Almost every one of its hundred small pages contains a sentence worth quoting. "The gyroscope that guides a rocket is an emissary from a six-dimensional symplectic world into our three-dimensional one; in its home world its behavior looks simple and natural." "Even those who see stars ask 'What is a star?,' because to see merely with one's eyes is still very little." "The image of Plato's cave seems to me the best metaphor for the structure of modern scientific knowledge; we actually see only the shadows." "In a world of light there are neither points nor moments of time; beings woven from light would live nowhere and nowhen; only poetry and mathematics are capable of speaking meaningfully about such things." "The screws and gears of the great machine of the world, when their behavior is understood, can be assembled and joined in a new order; thus one obtains a bow, a loom or an integrated circuit." "Modern theoretical physics is a luxuriant, totally Rabelaisian, vigorous world of ideas, and a mathematician can find in it everything to satiate himself except the order to which he is accustomed."

1. Translated from the Russian by Ann and Neil Koblitz (Birkhäuser, 1981).

Besides these verbal gems, Manin's book contains some equations and some technical exposition. His purpose was to make the thought processes of physics intelligible to mathematicians. He achieves this purpose by skillful selection of examples. Incidentally, by his style of writing and thinking, he makes the thought processes of a mathematician intelligible to physicists. He does not try to abolish or blur the distinction between mathematical and physical understanding. One of the many virtues of his book is that it leaves the central mystery, the miraculous effectiveness of mathematics as a tool for the understanding of nature, unexplained and unobscured.

Manin's book defies summary, because it already compresses into its small compass enough ideas to fill a dozen books of ordinary density. Each of its many topics is discussed without wastage of words. When I began trying to summarize it, I found myself willy-nilly selecting sentences and quoting them directly. The flavor of Manin's thinking is conveyed better by quotation than by paraphrase. I abandoned the attempt to describe the book in detail. Instead, I devote the rest of this review to the pursuit of a single question suggested by Manin's survey of contemporary science. Are we, or are we not, standing at the threshold of a new scientific revolution comparable with the historic revolutions of the past?

The two great conceptual revolutions of twentieth-century science, the overturning of classical physics by Werner Heisenberg and the overturning of the foundations of mathematics by Kurt Gödel, occurred within six years of each other within the narrow boundaries of German-speaking Europe. Manin sees no causal connection between the two revolutions. He describes them as occurring independently: "Physicists were disturbed by the interrelation between thought and reality, while mathematicians were disturbed by the interrelation between thought and formulas. Both of these relations turned out to be more complicated than had previously been thought, and the models, self-portraits and self-images of the two disciplines have turned out to be very dissimilar." This lucid characterization emphasizes the differences between the Heisenberg and Gödel revolutions. But a study of the historical background of German intellectual life in the 1920s reveals strong links between them. Physicists and mathematicians were exposed simultaneously to external influences that pushed them along parallel paths. Seen in the perspective of history, the geographical and temporal propinquity of Heisenberg and Gödel no longer appears to be a coincidence.

The historical dimension of science is explored in another short and excellent book, *Weimar Culture, Causality, and Quantum Theory, 1918–1927: Adaptation by German Physicists and Mathematicians to a Hostile Intellectual Environment*, by Paul Forman.[2] Forman is a historian, more familiar with physics than with mathematics. His book overlaps hardly at all with Manin's. To arrive at a balanced picture of our scientific heritage, the two books should be read together. I now turn my attention to Forman and come back to Manin later.

Forman begins with Felix Klein, sixty-nine years old and approaching the end of his long career as *grand seigneur* of German mathematics. It is June 1918, the last summer of World War I, and Klein is talking in Göttingen to an audience including leaders of German industry and of the Prussian government. He is addressing a formal session of the Göttingen Society for the Advancement of Applied Physics and Mathematics. He talks confidently of the coming victorious conclusion of the war, of the harmonious collaboration of German science with industry and the armed forces, and of the expected increase in support for mathematical education and research after the victory is won. Here in wartime Germany we see the first full flowering of the military-industrial complex in its modern style, soldiers and politicians sharing their dreams of glory with scientists and mathematicians. The Prussian minister of education responds to Klein with

2. University of Pennsylvania Press, 1971.

a generous grant of money for the foundation of a Mathematical Institute in Göttingen. Less than five months later, the dreams of glory have collapsed, the German Empire is utterly defeated, the Mathematical Institute indefinitely postponed. In the new era of defeat and misery that begins in November 1918, the exact sciences are discredited together with the military-industrial complex that had sustained them. The Göttingen Mathematical Institute is ultimately built, after Klein's death, not with German government funds but with American dollars supplied by the Rockefeller Foundation.

Forman uses Klein's Göttingen speech to set the stage for a dramatic description of the intellectual crosscurrents of Weimar Germany. The dominant mood of the new era was doom and gloom. The theme song was *Untergang des Abendlandes*, *Decline of the West*, the title of the apocalyptic world history of Oswald Spengler. The first volume of Spengler's prophetic work was published in Munich in July 1918, the month in which the tide of war on the western front finally turned against Germany. After the November collapse, the book took Germany by storm. It went through sixty editions in eight years. Everybody talked about it. Almost everybody read it. Forman demonstrates with ample documentation that mathematicians and physicists read it too. Even those who disagreed with Spengler were strongly influenced by his rhetoric. Spengler himself had been a student of science and mathematics before he became a historian. He had much to say about science. Not all of what he said was foolish. He said, among other things, that the decay of Western civilization must bring with it a collapse of the rigid structures of classical mathematics and physics. "Each culture has its own new possibilities of self-expression which arise, ripen, decay and never return. There is not one sculpture, one painting, one mathematics, one physics, but many, each in its deepest essence different from the other, each limited in duration and self-contained." "Western European physics—let no-one deceive himself—has reached the limit of its possibilities. This is the origin of the sudden and annihilating doubt that has arisen about things that even yesterday were the unchallenged foundation of physical theory, about the meaning of the energy principle, the concepts of mass, space, absolute time, and causal natural laws generally." "Today, in the sunset of the scientific epoch, in the stage of victorious skepsis, the clouds dissolve and the quiet landscape of the morning reappears in all distinctness.... Weary after its striving, the Western science returns to its spiritual home."

Two people who came early and strongly under the influence of Spengler's philosophy were the mathematician Hermann Weyl and the physicist Erwin Schrödinger. Both were writers with a deep feeling for the German language, and perhaps for that reason were easily seduced by Spengler's literary brilliance. Both became convinced that mathematics and physics had reached a state of crisis that left no road open except radical revolution. Weyl had been, even before 1918, a proponent of the doctrine of intuitionism, which denied the validity of a large part of classical mathematics and attempted to place what was left upon a foundation of intuition rather than formal logic. After 1918 he extended his revolutionary rhetoric from mathematics to physics, solemnly proclaiming the breakdown of the established order in both disciplines. In 1922 Schrödinger joined him in calling for radical reconstruction of the laws of physics. Weyl and Schrödinger agreed with Spengler that the coming revolution would sweep away the principle of physical causality. The erstwhile revolutionaries David Hilbert and Albert Einstein found themselves in the unaccustomed role of defenders of the status quo, Hilbert defending the primacy of formal logic in the foundations of mathematics, Einstein defending the primacy of causality in physics.

In the short run, Hilbert and Einstein were defeated and the Spenglerian ideology of revolution triumphed, both in physics and in mathematics. Heisenberg discovered the true limits of causality in atomic processes, and Gödel discovered the limits of formal deduction and proof in mathematics. And, as often happens in the history of intellectual revolutions, the achievement of revolutionary goals destroyed the revolutionary ideology that gave them birth. The visions of Spengler, having served their purpose, rapidly became irrelevant. The victorious revolutionaries were not irrational dreamers but rational scientists. The physics of Heisenberg, once it was understood, turned out to be as mundane and practical as the physics of Newton. Chemists who never heard of Spengler could successfully use quantum mechanics to calculate molecular binding energies. And in mathematics, the discoveries of Gödel did not lead to a victory of intuitionism but rather to a general recognition that no single scheme of mathematical foundations has a unique claim to legitimacy. After the revolutions were over, the new physics and the new mathematics became less and less concerned with ideology. In the long run, the value systems of physics and mathematics emerged from the revolutions essentially unchanged. Spengler's dream of a reborn, vitalistic, spiritualized science, "Western science returning to its spiritual home," was forgotten. The practical achievements of Hilbert and Einstein outlasted the fashionable despair of Spengler.

Now, fifty years later, the wheel has come full circle. The physics of quantum devices and the mathematics of effective computability have become everyday tools for engineers and industrialists to exploit. The new physics and the new mathematics are as friendly to the military-industrial complex of modern America as the old physics and the old mathematics were to the military-industrial complex of Germany in the days of Felix Klein. And once again we hear voices preaching revolution, a return to holistic thinking, a spiritualization of science. "Physics of Consciousness" is a fashionable slogan today, like the *Lebensphilosophie* of the 1920s. Fritjof Capra steps tentatively into the shoes of Oswald Spengler. Capra's *Tao of Physics*[3] is selling, like

3. Shambhala, 1975.

Spengler's *Untergang* of old, in hundreds of thousands of copies. Are we heading toward a period of radical changes in science, comparable with the Heisenberg revolution of 1925 and the Gödel revolution of 1931? Who can tell? Forman's historical analysis may illuminate the past, but it cannot predict the future.

Forman and Manin represent two contrasting styles in the historiography of science. Forman looks at science from the outside, Manin from the inside. Forman sees science responding to external social and political pressures; Manin sees science growing autonomously by the logical interplay of its own concepts. Forman takes his evidence from what scientists say, in speeches and writings directed toward the general public. Manin takes his evidence from what scientists do, as they exchange methods and ideas with one another. Forman is concerned with the rituals of science, Manin with the substance.

Looking back on the events of the 1920s with the benefit of hindsight, we can see clearly that Heisenberg and Gödel did not need Spengler to tell them what they had to do. It is true, as Forman demonstrates, that Spengler created a mood of revolutionary expectation in German-speaking Europe, and that the existence of this mood helps to explain why young people in Germany and Austria were better prepared than young people elsewhere to make revolutionary discoveries. But the discoveries of quantum mechanics and mathematical undecidability would have been made within a few years, either in German-speaking Europe or somewhere else, even if Spengler had never existed. The time was ripe for these discoveries, and the internal development of physics and mathematics made them inevitable. An external mood of revolutionary expectation is neither a necessary nor a sufficient condition for the occurrence of a scientific revolution. If we wish to assess realistically the prospects of scientific revolutions in the future, we should study science itself and not the philosophical or political ambience of science. We should leave Forman aside and go back to Manin.

The picture of present-day physics and mathematics that Manin presents to us is far removed from the intellectual turmoil of the 1920s. Manin's picture is idyllic. He shows us physics and mathematics as two neighboring gardens, each growing luxuriantly with trees and flowers in great variety, while the busy physicists and mathematicians fly to and fro like bees carrying pollen for the cross-fertilization of one plant by another. In Manin's gardens there is growth and decay, sunshine and showers, but no hint of gloom and doom. Looking to the future from Manin's perspective, one sees no evidence of a coming cataclysm, no sign of that "craving for crisis" which was, according to Forman, the hallmark of a German academic in the 1920s. On the contrary, Manin's picture of science promises us a long period of fruitful and multifarious growth, with plenty of surprises and sudden illuminations but no radical changes of objective. In Manin's view, the present epoch is characterized by a growing willingness of physicists and mathematicians to learn from one another and to transfer tools and techniques from one branch of science to another. The increasing overlap between physics and mathematics provides opportunities for the continuing enrichment of both disciplines. The future of physics and mathematics lies in evolution rather than revolution. Manin sees Spengler's "quiet landscape of the morning" not as an end but a beginning.

The concluding paragraph of Manin's book gives us a glimpse of his vision of the future:

> It is remarkable that the deepest ideas of number theory reveal a far-reaching resemblance to the ideas of modern theoretical physics. Like quantum mechanics, the theory of numbers furnishes completely non-obvious patterns of relationship between the continuous and the discrete, and emphasizes the role of hidden symmetries. One would like to hope that this resemblance is no accident, and that we are already hearing new words about

the world in which we live, but we do not yet understand their meaning.

Postscript, 2006

In the twenty-four years since this review was written, the development of mathematics and physics has continued as Manin predicted. The main area in which physicists and mathematicians are working together is string theory. Progress has been evolutionary rather than revolutionary. There is much talk of a radical revolution still to come, but no clear sign of its arrival.

15

EDWARD TELLER'S *MEMOIRS*

EDWARD TELLER'S *Memoirs: A Twentieth-Century Journey in Science and Politics*[1] is a pleasure to read and is also a unique historical document. Teller is intensely interested in people. The story of his life is a portrait gallery of people he has known, each of them brought to life and portrayed as an individual, all of them swept along by the tides of war and revolution and political passion in which Teller's life was lived. Teller observes and records the personal qualities of these people, their follies and their kindnesses and their often tragic fates, beginning with the friends of his childhood in Hungary eighty years ago and ending with the death of his wife, Mici, who loved and sustained him through more than seventy years of joys and sorrows.

Teller is also intensely interested in science. The high point of his life, as he describes it, was the brief golden age of German science, the seven years that he spent in Germany from 1926 to 1933, between the discovery of quantum mechanics and the advent of Hitler. During those years he worked on the boundary between physics and chemistry, understanding the implications of quantum mechanics for the structure and spectroscopy of molecules. He describes himself as a

1. Edward Teller with Judith Shoolery (Perseus, 2001). Teller died in 2003 at age ninety-five, only two years after these memoirs were published.

problem solver rather than a deep thinker. After the deep thinking had been done by Niels Bohr and Werner Heisenberg and Erwin Schrödinger, the road was open for problem solvers, such as Teller and his friends Hans Bethe and Lev Landau and George Gamow and Enrico Fermi, to apply the new ideas to practical problems. Using the new ideas, the problem solvers rebuilt physics and chemistry from the bottom up. Those seven years were indeed a golden age, when every young physicist could find important problems to solve, and when the number of physicists was so small that everyone knew everyone. Teller enjoyed the intense intellectual excitement of those years, and enjoyed even more the intense intellectual friendships. Like his friend Bethe, Teller was something of a poet. For a birthday party of Max Born in Göttingen, Teller composed a splendid song in German, with the rhythm and melody of the "Mack the Knife" tune from Brecht's *Threepenny Opera*. As a child, Teller was bilingual in Hungarian and German. Unfortunately, he says, he was eight years old before he began learning English, already too old to acquire the intimacy with words that a poet needs. After he moved to America and had to live his life in English, he stopped writing poems.

He sailed to America in 1935 on the same ship as Bethe, and taught physics at George Washington University while Bethe taught at Cornell. At GWU he was among old friends from Europe, Gamow and George Placzek and Maria Mayer. The first three years in America, from 1936 to 1938, were peaceful. Teller maintained his old friendships and made many new ones. The atmosphere of the golden age of German physics was almost recreated in America. Then, in December 1938, fission was discovered in Germany, and Teller's life was irreversibly changed. With Leo Szilard, another old friend from Hungary, he went to Einstein and persuaded Einstein to sign the famous letter that warned President Roosevelt of the possible military importance of fission. And from that time until today, Teller's life has been dominated by nuclear weapons. His experiences in Germany had burned into his soul the lesson that it was a fatal error for academic people to be unconcerned with the defense of freedom.

The second half of this book contains a detailed account of Teller's involvement with weaponry, first at Columbia, then in turn at Chicago, Los Alamos, and Livermore, and finally at Stanford. One might expect the narrative in this part of the book to become more political and less personal. But here too, even when Teller is most heavily engaged in political battles, he portrays his opponents as human beings and describes their concerns fairly. There is sadness in his account but no bitterness. The greatest sadness is the personal sadness, when three of his close friends and allies, Enrico Fermi, John von Neumann, and Ernest Lawrence, die untimely deaths before their work is done. Throughout his struggles he maintains his talent for friendship. Szilard, who disagreed violently with Teller about almost everything, remained one of his closest friends.

The worst period of Teller's life began in 1954 when he testified against J. Robert Oppenheimer in the hearing conducted by the Atomic Energy Commission to decide whether Oppenheimer was a security risk. The full transcript of Teller's testimony is included in the book. One result of Teller's testimony was that a large number of his friends ceased to be friends. The community of physicists that Teller loved was split apart. The hearing had been instigated by Oppenheimer's enemies in order to demonize him and destroy his political influence. After the hearing, it was Teller's turn to be demonized. Oppenheimer and Teller both suffered grievously from the quarrel, but the damage to Teller was greater. I remember meeting Bethe in Washington while the hearing was in progress, shortly before Teller testified. Bethe was looking grimmer than I had ever seen him. He said, "I have just now had the most unpleasant conversation of my whole life. With Edward Teller." Bethe had tried to persuade Teller not to testify and had failed. That was the end of a twenty-year friendship. Bethe and Teller are now the last survivors of the golden age. I was happy to read in *Physics Today* a review of this book by Bethe, a generous review, emphasizing the warmth of Teller's character and letting old quarrels sleep.

Teller's account of his testimony has been challenged by the historian Gregg Herken in a less generous review of this book in *Science*. Herken emphasizes some details in Teller's account that disagree with historical documents. But a historian should be familiar with the fact that all human memories of past events are unreliable. Memoirs are not history. They are the raw material of history. Memoirs written by generals and politicians are notoriously inaccurate. When I wrote my own memoirs some years ago, I was amazed to discover how many things I remembered that never happened. Memory not only distorts but also invents. A writer of memoirs should make an honest effort to set down the course of events as they are recorded in memory. This Teller has done. If some of the details are wrong, that detracts little from the value of this book as a panorama of a historical epoch in which Teller played a leading role. His account of his testimony in the Oppenheimer hearing ends with the statement, "I proved not only that stupidity is a general human property but that I possessed a full share of it." When Oppenheimer was asked by his interrogator during the hearing why he had lied to security officers, he replied, "Because I was an idiot." Teller is saying that he was an idiot too, when he voluntarily agreed to take part in a dirty business. That is Teller's conclusion, and it is a fair summary of his role in the affair.

Teller was not only the main inventor of the hydrogen bomb but also the main driving force pushing its development. For this he makes no apology. He believes that United States' possession of the hydrogen bomb was essential to the peaceful resolution of the cold war. But he also writes admiringly of Andrei Sakharov, who pushed the development of hydrogen bombs in the Soviet Union for similar reasons. Hydrogen bombs on both sides of the cold war were essential to keeping it cold. One evening during the 1960s, I was drinking beer in Germany with a German friend who had spent most of the Second World War as an infantry officer in Russia. He talked eloquently of the joys of the Russian campaign, how civilian life was petty and boring compared with the heroism he had witnessed in Russia, how his years as a soldier in Russia were the best years of his life. Then he pointed a finger at me and said, "If it were not for your damned hydrogen bombs, we would be back in Russia today." At that moment I was thinking, "Thank God for Edward Teller and his bombs."

Some of the most illuminating passages in the book are extracts from letters written by Teller to Maria Mayer. Mayer was a first-rate physicist and also the friend to whom Teller confided his feelings at moments of maximum stress. Here is a passage written in early 1950, when Teller was engaged at Los Alamos in his lonely struggle to build a hydrogen bomb, a year before the crucial invention that made the bomb possible: "Whatever help and whatever advice I can get from you—I need it. Not because I feel subjectively that I must have help, but because I know objectively that we are in a situation in which any sane person must and does throw up his hands and only the crazy ones keep going."

Another illuminating passage is a quote from a letter written in 1939 by Merle Tuve, a senior physicist who knew Teller during his years at George Washington University. Somebody at the University of Chicago had asked Tuve for an appraisal of Teller. Tuve replied, "If you want a genius for your staff, don't take Teller, get Gamow. But geniuses are a dime a dozen. Teller is something much better. He helps everybody. He works on everybody's problem. He never gets into controversies or has trouble with anyone. He is by far your best choice." That was the Teller I knew when I worked with him for three months in 1956 on the design of a safe nuclear reactor. It was easy to disagree fiercely about the details of the reactor, as we often did, and remain friends. He helped everybody and worked on everybody's problem. There was of course another Teller, the Teller who worked crazily for unpopular causes such as hydrogen bombs and missile defense, and who fought furiously for the causes that he believed in. This book gives us a fair portrait of both Tellers, the Teller who gave generous help to young scientists and the Teller who quarreled vehemently with older scientists. Those who disagreed with him did him a grave injustice when they tried to turn him into a demon.

16

IN PRAISE OF AMATEURS

TIMOTHY FERRIS IS a serious amateur astronomer. He spends a substantial amount of his time and money roaming around at night among planets and stars and galaxies. He owns a place called Rocky Hill Observatory in California where he can stargaze to his heart's content through telescopes of modest size and excellent quality. He belongs to the international community of observers who are linked by the Internet as well as by the shared sky in which they are at home. Serious amateur astronomers, unless they are retired or independently wealthy, must have a day job to support their nocturnal addiction. Ferris has a day job as a writer of books explaining science to the general public. He has written many books which are widely read and have effectively reduced the level of scientific illiteracy of the American population.

Seeing in the Dark: How Backyard Stargazers Are Probing Deep Space and Guarding Earth from Interplanetary Peril[1] is similar to the others in some respects and different in others. Like his previous books, it is factually accurate, it contains a wealth of information about the universe we live in, and it makes the information easily digestible by seasoning it with good stories. Unlike his other books, it

1. Simon and Schuster, 2002.

is a love story, describing how Ferris fell in love with astronomy at the age of nine and how this passion has enriched his life ever since. But he does not write much about himself. The book is mainly a portrait gallery of the diverse and colorful characters who have shared his passion, with a description of the contributions that they have made to the science of astronomy.

Ferris has sought out his amateur astronomical colleagues, visited them in their homes and observatories, listened to their life stories, and watched them at work. One of these colleagues is Patrick Moore, who has also supported himself by writing popular science books in the daytime while exploring the sky at night. Ferris visited him in the English village of Selsey where he lives and works. Many years ago, before any human beings or human instruments had surveyed the back side of the moon from space, Moore was observing the moon systematically with his small telescope at Selsey.

The moon normally keeps a fixed orientation as it revolves around the Earth, so that only the front side is visible. But it wobbles slightly in its orbit, so that occasionally some regions that are normally invisible can be seen at the edge of the visible face, extremely foreshortened and inconspicuous. Moore was studying these normally invisible regions at a moment when the moon's wobble was at a maximum, and discovered Mare Orientale, the biggest and most beautiful impact crater on the moon. Moore gave it the name Mare Orientale, Eastern Sea, because it is hidden behind the eastern edge of the moon and because it is a dark circular region similar to the dark regions on the front side of the moon which the amateur astronomer Johannes Hevelius called seas when he mapped them in 1647.

Hevelius was a brewer in Danzig who made the first accurate map of the moon. Even at times of maximum wobble, only a small part of Mare Orientale can be seen from the Earth. Only an observer with long experience and deep knowledge of lunar topography could have recognized it in the fragmentary view of the moon's edge that Moore could see from Selsey. Professional astronomers do not have such experience or such knowledge. Only an amateur could have discovered Mare Orientale, because only an amateur has the time and the motivation to study a single region of the moon with single-minded dedication.

Patrick Moore is one of many examples illustrating the main theme of Ferris's book. The theme is the importance of amateurs in the exploration of the universe, not only in past centuries but also today. Moore was an old-fashioned amateur when he discovered Mare Orientale, observing the moon laboriously with his eye at the telescope, drawing maps of his observations with pencil and paper. Amateurs today observe the sky with digital electronic cameras, recording the images with personal computers using commercial software. The role of amateurs has become more important in the last twenty years because of the advent of cheap mass-produced electronic cameras, computers, and software. Serious amateurs today can afford to own equipment that few professional observatories could afford twenty years ago. Personal computers are used not only to record data but to communicate rapidly with other observers and to coordinate observations all over the world.

There are many areas of research that only professional astronomers can pursue, studying faint objects far away in the depths of space, using large telescopes that cost hundreds of millions of dollars to build and operate. Only professionals can reach halfway back to the beginning of time, to explore the early universe as it was when galaxies were young and the oldest stars were being born. Only professionals have access to telescopes in space that can detect the X-rays emitted by matter heated to extreme temperatures as it falls into black holes.

But there are other areas of research in which a network of well-equipped and well-coordinated amateurs can do at least as well as the professionals. Amateurs have two great advantages, the ability to survey large areas of sky repeatedly and the ability to sustain observations over long periods of time. As a result of these advantages, amateurs are frequently first to discover unpredictable events such as storms in the atmospheres of planets and catastrophic explosions of stars. They compete with professionals in discovering transient objects such as comets and asteroids. It often happens that an amateur makes a discovery which a professional follows up with more detailed observation or theoretical analysis, and the results are then published in a professional journal with the amateur and the professional as co-authors.

On Mount Palomar in California there are two famous telescopes, the huge 200-inch and the little 18-inch. The 200-inch was for many years the largest in the world, exploring the far reaches of the universe with unequaled sensitivity. The 18-inch was on the mountain before the 200-inch and made equally important discoveries. It was the brainchild of the German amateur astronomer Bernhardt Schmidt. Schmidt was a professional optician who made a living by grinding lenses and mirrors. He worked as an unpaid guest at the university observatory in Hamburg. In 1929 he invented a new design for a telescope that produced sharply focused photographic images over a wide field of view. He built and installed the first Schmidt telescope at Hamburg. The Schmidt telescope made it possible for the first time to photograph large areas of sky rapidly. Compared with previously existing telescopes, the Schmidt could photograph about a hundred times more area every night.

The 18-inch at Palomar was the second Schmidt telescope to be built and the first to be used in a mountaintop observatory with good astronomical seeing. Fritz Zwicky, a Swiss professional astronomer, understood the potential of Schmidt's invention and installed the 18-inch on the mountain in 1935. He used it to do the first rapid photographic sky survey, photographing large areas of sky every night and mapping the positions of hundreds of thousands of galaxies. As a result of this survey, Zwicky made two fundamental discoveries. He found that galaxies have a universal tendency to congregate into clusters. And he found that the visible mass of the galaxies is insufficient to account for the clustering. From the observed positions and velocities of the galaxies, Zwicky calculated that the clusters must contain invisible mass that is about ten times larger than the visible mass. His discovery of the invisible mass, made with the little Schmidt telescope, opened a new chapter in the history of cosmology. Our later explorations of the cosmos have confirmed that Zwicky was right, that the dark unseen mass dominates the dynamics of the universe. Professional and amateur astronomers are using Schmidt telescopes all over the world to continue the revolution that Schmidt and Zwicky started. Schmidt himself did not live to see the triumph of his invention. When Hitler came to power in Germany in 1933, Schmidt was so disgusted that he gave up hope and quietly drank himself to death.

David Levy is an amateur astronomer in the modern style. He observes at his home in Arizona where he has three modest but well-equipped telescopes, two of them of Schmidt design. He has also visited frequently as a guest observer at the Palomar observatory in California, where he collaborates with the professionals. At Palomar he was using Zwicky's original 18-inch telescope, which was still going strong and making important discoveries after sixty years of intensive use. Levy's collaborators were Eugene and Carolyn Shoemaker, until Eugene's untimely death in a car accident. Now he continues the collaboration with Carolyn alone.

The most famous event of the collaboration occurred in 1993 when Eugene was still alive. This was the discovery of the comet Shoemaker-Levy 9, which was seen in the process of tidal disruption after passing too close to the planet Jupiter. The newly discovered comet was at that moment breaking up into eighteen pieces. The pieces moved apart until they looked like a string of pearls, stretched out into a straight line, each with its own tail of gas and dust shining in the light of the distant sun. After a few days of careful observation and calculation, it became clear that the pieces of the comet were all destined to crash into Jupiter sixteen months later. This was the first time in the history of astronomy that two celestial objects were seen to collide.

At the time when Jupiter was under bombardment in July 1994, I was lucky to be a guest of the amateur astronomer Gilbert Clark in the dome occupied by a 24-inch telescope on Mount Wilson in California. Clark is a retired navy officer who founded and directs a charitable foundation called Telescopes in Education, or TIE for short. The 24-inch telescope is on loan from the Mount Wilson Observatory to TIE and is instrumented so that it can be operated by remote control. While Clark and I were in the dome, the telescope was being operated by children in a classroom in Virginia. We could see the same images that the children were seeing, and we could hear their voices. They were deciding where to point the telescope. They looked intermittently at various deep-sky objects, galaxies, and star clusters, but always came back to Jupiter. There on the screen was Jupiter, not the familiar image of Jupiter with bland horizontal bands in its cloudy atmosphere, but a wounded Jupiter with five big black scars at the places where fragments of the comet had struck. To me the most remarkable feature of the view was that we could see Jupiter spinning. The scars made the rotation of the planet visible. Jupiter spins fast, making one revolution in nine hours, forty degrees of longitude per hour. We could see the scars moving across the face of the planet, disappearing at one edge and appearing at the other. And the children could see them too.

Ferris is saying that amateur astronomy is a growth industry, gaining in scientific importance as new technologies increase the reach of amateur instruments. Another factor favoring the amateur observer is the change in our view of the universe caused by recent discoveries. The traditional Aristotelian view imagined the astronomical universe to be a sphere of unchanging peace and harmony. The earth alone was perishable and violent, while the heavenly bodies were perfect and quiescent. This view was contradicted by a multitude of discoveries during the last four hundred years, beginning with the two exploding stars observed by Tycho Brahe and Johannes Kepler and with the mountains and valleys discovered by Galileo on the moon. In the last fifty years it became clear that we live in a violent universe, full of explosions, collapses, and collisions. The Earth now appears to be a comparatively quiet corner in a universe of cosmic mayhem. The 1994 bombardment of Jupiter demonstrated that our own solar system is not immune to cosmic violence. After this replacement of the old static view of the universe by a new dynamic view, the subject matter of astronomy is also transformed. Astronomy is less concerned with things that do not change and more concerned with things that change rapidly. The new emphasis on rapidly changing phenomena requires quick and frequent observation. Quick and frequent observation is a game that serious amateurs can play well. It is a game that amateurs can sometimes play better than professionals. It is a game that gives amateurs and professionals many opportunities for fruitful cooperation.

Ferris shows us a grand vision of the growing importance of amateurs, nimble, well-equipped, and well-coordinated, jumping ahead of the slow-moving professionals to open new frontiers. Some professional astronomers share this vision and welcome the help that amateurs can provide. But most professionals consider the efforts of the amateurs trivial. After all, the professionals with their big instruments and big projects are solving the central problems of cosmology, while the amateurs are finding pretty little comets and asteroids. The view of the majority of professionals was expressed by the physicist Ernest Rutherford, the discoverer of the atomic nucleus, who said: "Physics is the only real science, the rest is butterfly-collecting." For most professional astronomers, the large-scale structure of the universe is real science, while comets and asteroids are unimportant details of interest only to butterfly collectors. Butterfly-collecting is an amiable hobby, but it should not be confused with serious science.

The clash between the two visions of amateur astronomy, Ferris's vision of amateurs as pioneer explorers and Rutherford's vision of amateurs as butterfly collectors, has deep roots. It arises from an ancient clash between two visions of the nature of science. There are two kinds of science, known to historians as Baconian and Cartesian. Baconian science is interested in details, Cartesian science is interested in ideas. Bacon said:

> All depends on keeping the eye steadily fixed on the facts of nature, and so receiving their images as they are. For God forbid that we should give out a dream of our own imagination for a pattern of the world.

Descartes said:

> I showed what the laws of nature were, and without basing my arguments on any principle other than the infinite perfections of God I tried to demonstrate all those laws about which we could have any doubt, and to show that they are such that, even if God created many worlds, there could not be any in which they failed to be observed.

Modern science leapt ahead in the seventeenth century as a result of fruitful competition between Baconian and Cartesian viewpoints. The relation between Baconian science and Cartesian science is complementary. We need Baconian scientists to explore the universe and find out what is there to be explained. We need Cartesian scientists to explain and unify what we have found. Generally speaking, professional astronomers tend to be Cartesian, amateur astronomers to be Baconian. It is right and healthy that there should be a clash between their viewpoints, but it is wrong for either side to treat the other with contempt. Ferris's sympathies are on the side of the amateurs, but he portrays the professionals with respect and understanding.

Astronomy is the oldest science and has the longest history. For two thousand years it was studied in different ways in two disconnected worlds, the Western world of Babylonia and Greece and Arabia, and the Eastern world of China and Korea. Ancient astronomy in the West was predominantly Cartesian, culminating in the elaborate theoretical universe of Ptolemy, with the clockwork machinery of cycles and epicycles determining how the heavenly bodies should move. Astronomy in the East was Baconian, collecting and recording observations without any unifying theory. In both worlds, astronomy was mixed up with astrology and was mainly studied by professional astrologers. After a promising start, progress stopped and science stagnated for a thousand years, because neither Baconian science nor Cartesian science could flourish in isolation from each other. In the West, theory was unconstrained by new observations, and in the East, observations were unguided by theory.

Then came the great awakening in the West, when Bacon and Descartes together led the way to the flowering of modern science. The seventeenth and eighteenth centuries were the heyday of the scientific amateurs. During those two centuries, professional scientists like Isaac Newton were the exception and gentleman amateurs like his rival Gottfried Leibniz were the rule. Amateurs had the freedom to jump from one area of science to another and start new enterprises without waiting for official approval. But in the nineteenth century, after two hundred years of amateur leadership, science became increasingly professional. Among the leading scientists of the nineteenth century, professionals such as Michael Faraday and James Clerk Maxwell were the rule and amateurs Charles Darwin and Gregor Mendel were the exceptions. In the twentieth century the ascendancy of the professionals became even more complete. No twentieth-century amateur could stand like Darwin in the front rank with Edwin Hubble and Albert Einstein.

If Ferris is right, astronomy is now moving into a new era of youthful exuberance in which amateurs will again have an important share of the action. It appears that each science goes through three phases of development. The first phase is Baconian, with scientists exploring the world to find out what is there. In this phase, amateurs and butterfly collectors are in the ascendant. The second phase is Cartesian, with scientists making precise measurements and building quantitative theories. In this phase, professionals and specialists are in the ascendant. The third phase is a mixture of Baconian and Cartesian, with amateurs and professionals alike empowered by the plethora of new technical tools arising from the second phase. In the third phase, cheap and powerful tools give scientists of all kinds freedom to explore and explain. The most important of the new tools is the personal computer, now universally accessible and giving amateurs the ability to do quantitative science. After the computer, the next-most-important tool is the World Wide Web, giving amateurs access to scientific papers and discussions before they are published, allowing amateurs all over the world to communicate and work together.

Astronomy, the oldest science, was the first to pass through the first and second phases and emerge into the third. Which science will be next? Which other science is now ripe for a revolution giving opportunities for the next generation of amateurs to make important discoveries? Physics and chemistry are still in the second phase. It is difficult to imagine an amateur physicist or chemist at the present time making a major contribution to science. Before physics or chemistry can enter the third phase, these sciences must be transformed by radically new discoveries and new tools. The status of biology is less clear. Mainstream biology is undoubtedly in the second phase, dominated by armies of professionals exploring genomes and analyzing metabolic pathways. But there is a wide hinterland of biology away from the mainstream, where amateurs following the tradition of Darwin discover new species of wildflowers, breed new varieties of dogs and pigeons and orchids, and collect butterflies. The writer Vladimir Nabokov is the most famous of twentieth-century butterfly collectors, but there are many others not so famous who also discovered new species. A young friend of mine who went recently as a student to Ecuador discovered twelve new species of plants in the rain forest.

Biology will probably be the next science to enter the third stage. New tools which might give power to amateur biologists are already visible on the horizon. The new tools will be cheaper and smaller versions of the tools now used by professional biologists to do genetic engineering. It took thirty years for the expensive and cumbersome mainframe computers of the 1950s to evolve into the cheap and convenient personal computers of the 1980s. In a similar fashion, the expensive genome-sequencing and protein-synthesizing machines of today will evolve into cheap machines that can stand on a desktop. The personal computer is not only cheaper and smaller, but also faster and more powerful than the mainframe that it replaced. The desktop sequencers and synthesizers of the future will be faster and more powerful than the machines that they will replace, and will be controlled by more sophisticated computer programs.

When these tools are available, the demand for them will be irresistible, just as the demand for laptop computers is irresistible today. Genetic engineering of roses and orchids, ornamental shrubs and vegetables, will be a new art form as well as a new science. Homeowners in well-to-do suburbs will use the new tools to embellish their gardens, while subsistence farmers in poor countries will use them to feed their families with higher-yielding or better-tasting potatoes. Amateur plant breeders and animal breeders and ecologists and nature lovers will then be enabled to make serious contributions to science, just as amateur astronomers do today.

Before the amateur use of genetic engineering becomes widespread, numerous political and legal obstacles will have to be overcome. Many people are strongly opposed to genetic engineering of any kind. Some of the opposition arises from religious or ideological principles, but much of it arises from practical concerns. Genetic engineering can undoubtedly be dangerous to public health and to ecological stability. The use of genetic engineering kits must be strictly regulated if these dangers are to be avoided. Genetic engineering of microbes is a great tool for terrorists, as Richard Preston demonstrates in his recent book *The Demon in the Freezer*.[2] Any kit available to the public must be made physically incapable of handling microbes. It could well happen that political authorities will decide to prohibit such kits altogether. It will be a sad day for biology if amateurs are forbidden the use of tools available to professionals. But that is a decision which we should leave to our grandchildren.[3]

When we look at the wider society outside the domain of science, we see amateurs playing essential roles in almost every field of human activity. Amateur musicians create the culture in which professional musicians can flourish. Amateur athletes, amateur actors, and amateur environmentalists improve the quality of life for themselves and others. Amateur writers such as Jane Austen and Samuel Pepys do as much as the professionals Charles Dickens and Fyodor Dostoevsky to plumb the heights and depths of human experience. In the most important of all human responsibilities, the raising of children and grandchildren, amateurs do the lion's share of the work. In almost all the varied walks of life, amateurs have more freedom to experiment and innovate. The fraction of the population who are amateurs is a good measure of the freedom of a society. Ferris shows us how

2. Random House, 2002.

3. The theme of amateur biology is explored further in my forthcoming book, *A Many-Colored Glass: Reflections on the Place of Life in the Universe* (University of Virginia Press, 2006).

amateurs are giving a new flavor to modern astronomy. We may hope that amateurs in the coming century, using the new tools that modern technology is placing in their hands, will invade and rejuvenate all of science.

17

A NEW NEWTON

IT WAS A strange juxtaposition. A big metal box filled with the manuscripts of Isaac Newton, hidden by Newton during his lifetime and unread for two hundred years afterward, and a fat young man with red hair and khaki shorts, strutting on the stage at meetings of the British Union of Fascists. The big metal box was packed up by Newton in 1696, when he left Cambridge and moved to London. He was leaving forever the life of intense and solitary study that he had pursued in Cambridge for thirty-five years, and entering the role of public figure and patron saint of the Age of Enlightenment that he pursued in London for thirty years more. The fat young man was Lord Lymington, Earl of Portsmouth. He was a direct descendant of Catherine Barton, the niece of Newton who kept house for him in London and inherited his papers when he died. Catherine Barton's daughter Kitty married an Earl of Portsmouth and became an ancestor of the fat young man. And so the fat young man came into possession of the big metal box. When he came into possession of the box, the papers inside were still intact.

When I was a boy in high school during World War II, I met the fat young man and disliked him intensely. I was helping England to survive by bringing in the harvest, at a time when the grown-ups who normally worked on the farms had been called up to serve in the army. The high school kids worked hard in the fields and enjoyed taking a holiday from Latin and mathematics. But the fat young man owned the land where we were working, and he came and lectured us about blood and soil and the mystical virtues of the open-air life. He had visited Germany, where his friend Adolf Hitler had organized the schoolkids to work on the land in a movement that he called *Kraft durch Freude*, in English "Strength through Joy." In Germany the kids had an accordionneuse, a woman with an accordion who played music to them all day long and kept them working in the right rhythm. The fat young man said he would find an accordionneuse for us too. Then we would have strength through joy and we would be able to work much better. Fortunately the accordionneuse never showed up, and we continued to work in our own rhythm. We knew that the fat young man was second in command to Sir Oswald Moseley in the British Union of Fascists, and if his friend Adolf had successfully invaded England he would probably have been our *Gauleiter*. Being well-brought-up English children, we listened to the fat young man politely and never showed him our contempt.

When I was bringing in the harvest and listening to the fat young man, I did not know that he had been the owner of the Newton papers. I learned this two years later from the economist John Maynard Keynes. I was then a student at Trinity College, Cambridge, while Keynes was a fellow of King's College. Keynes was chief economic adviser to the British government and largely responsible for keeping the British economy afloat at a time when more than half of our gross national product, and all of our foreign exchange, was being spent on the war. He was wearing himself out, flying back and forth between London and Washington and dealing with one financial crisis after another. He never had time to pursue his hobby, the careful scholarly reading of the Newton papers. I was lucky to be present at one of his rare appearances in Cambridge, when he gave a lecture at Trinity College with the title "Newton, the Man."[1] The audience was small, and we huddled around the exhausted figure of Keynes as he lay in a reclining chair in a cold, dark room and talked quietly about the big metal box and its contents. Four years later he died of heart failure, precipitated by overwork and the hardships of crossing the Atlantic repeatedly in slow propeller-driven airplanes under wartime conditions.

Keynes described to us how the fat young man, in need of cash to finance the British Union of Fascists in 1936, had brought the big metal box to Sotheby's in London and sold the contents at auction in 329 separate lots. Keynes had warning of the sale only a few days before it happened. He attended the auction and bought as many of the papers as he could with money out of his own pocket. "Disturbed by this impiety," he told us, "I managed gradually to reassemble about half of them, including nearly the whole of the biographical portion, in order to bring them to Cambridge which I hope they will never leave. The greater part of the rest were snatched out of my reach by a syndicate which hoped to sell them at a high price, probably in America." The papers that he rescued are now preserved in the King's College library. The rest of them were sold piecemeal to various collectors and dispersed all over the world. Even as a salesman of irreplaceable antiquities, the fat young man was incompetent. As a reward for his act of gross impiety, he reaped a total of only £9,000.

In his lecture, Keynes described the contents of the box that he examined as best he could during the turmoil of the sale and afterward. Among the papers that he rescued was a firsthand description of Newton, written by Newton's cousin Humphrey, who worked for him as a secretary for five years. Those five years included the two

---

1. John M. Keynes, "Newton, the Man," in *Newton Tercentenary Celebrations* (Royal Society of London, Cambridge University Press, 1947), pp. 27–34. Since Keynes died in 1946, his lecture was read at the tercentenary celebrations by his brother Geoffrey Keynes.

that Newton spent writing his *Philosophiae Naturalis Principia Mathematica*, in English *Mathematical Principles of Natural Philosophy*, the masterpiece that set the course of the physical sciences for the next two hundred years. The *Principia* appeared in three volumes. The first two established the laws of physics and the methods of calculating the consequences of the laws. The third volume begins with Newton's proud statement: "It remains that, from the same principles, I now demonstrate the frame of the system of the world." The third volume analyzes the diverse phenomena of the real world, the motions of sun and moon, planets, satellites, and comets, the precession of the earth's axis of rotation, and the rise and fall of tides, and shows how they all occur precisely as his principles predict. The manuscript of the *Principia*, which Newton's friend Edmond Halley took with him to London in 1686 to be published, is in Humphrey Newton's hand.

Humphrey's description of Newton's life in Cambridge was written many years later. Newton spent much of his time in the elaboratory, a wooden building in his garden in which he did alchemical experiments. Here is Humphrey writing about Newton as an alchemist:

> Especially at spring and fall of the leaf, at which times he used to imploy about six weeks in his Elaboratory, the fire scarcely going out either night or day, he sitting up one night, as I did another, till he had finished his chymical experiments, in the performances of which he was the most accurate, strict, exact. What his aim might be, I was not able to penetrate into, but his pains, his diligence at those set times, made me think, he aimed at something beyond the reach of human art and industry.

In his talk at Trinity College, Keynes quoted Humphrey and then added his own interpretation:

> Newton was clearly an unbridled addict.... He was almost

> entirely concerned, not in serious experiment, but in trying to read the riddle of tradition, to find meaning in cryptic verses, to imitate the alleged but largely imaginary experiments of the initiates of past centuries. Newton has left behind him a vast mass of records of these studies. I believe that the greater part are translations and copies made by him of existing books and manuscripts. But there are also extensive records of experiments.... In these mixed and extraordinary studies, with one foot in the Middle Ages, and one foot treading a path for modern science, Newton spent the first phase of his life, the period of life in Trinity when he did all his real work.... And when the turn in his life came and he put his books of magic back into the box, it was easy for him to drop the seventeenth century behind him and to evolve into the eighteenth-century figure which is the traditional Newton.... And he looked very seldom, I expect, into the chest where, when he left Cambridge, he had packed all the evidences of what had occupied and so absorbed his intense and flaming spirit in his rooms and his garden and the elaboratory between the Great Gate and Chapel.

During the sixty years since Keynes spoke in Cambridge, the papers that were hidden in the big metal box have given rise to a literature that is even more voluminous. The collected mathematical papers of Newton have been published in eight large volumes, the collected correspondence in seven. The standard biography of Newton by Richard Westfall, with the title *Never at Rest*,[2] fills more than nine hundred pages. Large numbers of more specialized books and papers have been devoted to Newton's mathematics, optics, physics, alchemy, and theology, to his scientific quarrels, his religious beliefs, and his later official career as Master of the Mint.

---

2. Cambridge University Press, 1983.

Now comes a new biography by James Gleick.[3] For the casual reader with a serious interest in Newton's life and work, I recommend Gleick's biography as an excellent place to start. It has three important virtues. It is accurate, it is readable, and it is short. It is roughly one quarter of the length of Westfall's book, and still gives a well-rounded and fairly complete picture of Newton and his ideas. To take the subject of alchemy as an example, Newton's alchemical activities occupy forty-six pages of Westfall (half each of Chapters 8 and 9), eight pages of Gleick (Chapter 9 with the title "All Things Are Corruptible"). Gleick's account is more sharply focused on the essential question, how it was possible for a mind as sharp and logical as Newton's to search for nature's secrets in ancient alchemical manipulations as well as in physical laws. Gleick's answer to this question:

> It was God who breathed life into matter and inspired its many textures and processes.... Rather than turn away from what he could not explain, he plunged in more deeply.... There were forces in nature that he would not be able to understand mechanically, in terms of colliding billiard balls or swirling vortices. They were vital, vegetable, sexual forces—invisible forces of spirit and attraction. Later, it had been Newton, more than any other philosopher, who effectively purged science of the need to resort to such mystical qualities. For now, he needed them.

During the years after the auction, two other scholarly collectors besides Keynes were slowly reassembling the papers that were scattered in 1936. The other two were Roger Ward Babson, an American stock market analyst, and A. S. Yahuda, an Orientalist born in the Middle East who ended up at Yale University. It was fortunate that these three collectors had interests that did not strongly overlap. Keynes

3. *Isaac Newton* (Pantheon, 2003).

was primarily interested in papers concerned with alchemy, Babson in papers concerned with gravitation, Yahuda in papers concerned with theology. The Babson collection is now in the Babson College Library at Wellesley, Massachusetts; the Yahuda collection is in the Jewish National and University Library in Jerusalem. Contrary to Keynes's fears, the papers that went to America are in collections open to scholars, while the few papers that remain inaccessible are mostly in France and Switzerland.

The Yahuda collection gives us an intimate view of Newton's religious thinking, which was as intense and idiosyncratic as his thinking about alchemy and mathematical physics. He saw clearly that there is no firm basis in scripture for the orthodox Christian doctrine of the Trinity. He was a Unitarian, deducing from the evidence of scripture that God the Father reigns alone. There is one God and not three. Jesus is his son and the Holy Spirit is his mouthpiece, but neither of them is his equal. All through his life, Newton was searching for truth in ancient writings as well as in the study of nature. He considered his Unitarian theology to be as firmly based as his mathematical physics.

But there was a practical difference between physics and theology. He was free to say whatever he liked about physics, but not about theology. Cambridge University and Trinity College were religious foundations with strict standards of orthodoxy. Newton could not have held his positions as professor at the university and fellow of the college if his heretical views had been publicly known. Fortunately, King Charles II, a man of liberal temperament, signed a special dispensation that excused Newton from the usual rule that university professors must be priests of the Anglican Church. To become a priest, Newton would have had to affirm his belief in the orthodox Trinitarian doctrine of the Church, and this he could never have done. In effect, the King was adopting a policy of "Don't ask, don't tell," and Newton carried out his side of the bargain by keeping his theological writings hidden in the big metal box.

Gleick describes Newton's theology in an excellent short chapter with the title "Heresy, Blasphemy, Idolatry," but he does not share Newton's enthusiasm for the fine points of biblical scholarship. He quotes with approval Westfall's judgment that *The Chronology of Ancient Kingdoms Amended*, a book written by Newton in his old age and published after his death, is "a work of colossal tedium." Anyone who would like a more sympathetic and more detailed account of Newton's religious studies, based on the Yahuda papers in Jerusalem, should read the book *The Religion of Isaac Newton* by Frank Manuel.[4] Manuel's book is, so far as I know, the only important work about Newton that does not appear in Gleick's bibliography.

For several years after the publication of the *Principia* in 1687, Newton was deeply involved in national politics. The "Glorious Revolution" of 1688 was a turning point in English constitutional history, as important for England as the revolution of 1776 was for America. In 1688 the country rose in rebellion against King James II, who stood for the divine right of kings, and drove him into exile. King William III was invited to come over from Holland to take his place. The essential part of the deal was that William would be a constitutional monarch, subject to the law of the land as determined by the English Parliament.

When James II provoked the constitutional crisis of 1687, Newton was a member of Parliament representing Cambridge University. The independence of the university was directly threatened by the King, who took action to remove Protestants and install Catholics in the university administration. Newton took a firm line against the King. "Be courageous therefore and steady to the laws," he wrote in a memorandum to the university. "If one P[apist] be a Master you may have a hundred.... An honest Courage in these matters will secure

4. Oxford University Press, 1974.

all, having Law on our sides." After successfully resisting King James, Newton urged the university to accept allegiance to King William so long as King William upheld the law of the land. In a letter to a friend in 1689, Newton described the agreement that the leaders of the university had accepted. In this letter he expressed, with his usual clarity, the fundamental principles of constitutional government:

> 1. Fidelity & Allegiance sworn to the King, is only such a Fidelity & Obedience as is due to him by the law of the Land. For were that Faith and Allegiance more than what the law requires, we should swear ourselves slaves & the King absolute, whereas by the Law we are Free men notwithstanding those oaths. 2. When therefore the obligation by the law to Fidelity and Allegiance ceases, that by the oath also ceases.

Sarah Jones Nelson, a colleague of mine in Princeton, recently discovered in the archives of Magdalen College, Oxford, another document either in Newton's hand or in that of a scribe (who was hired for the purpose of working quickly). It was put into the archive by the philologist R. W. Chapman, who had bought it in the auction at Sotheby's in 1936, but nobody else seems to have been aware of its existence. Internal evidence shows that it was written in 1687 or 1688. It outlines the legal case against King James II, and also suggests the relationships between scientific knowledge, law, and morality.

It appears that at that time Newton was searching for a common foundation for physical law and moral law, seeing both kinds of law as manifestations of the same divine wisdom. While he was attending the sessions of Parliament in London, he met the philosopher John Locke, the great protagonist of government by consent of the governed. Locke shared his interests, in theology as well as in politics. He was, like Newton, a closet Unitarian. In a letter to another friend, Locke remarks, "Mr. Newton is really a very valuable man, not only for his wonderful skill in mathematics, but in divinity too, and his great knowledge in the Scriptures, wherein I know few his equals." According to Sarah Jones Nelson, the Magdalen manuscript contains ideas concerning the moral and legal theory of civil disobedience, which reappear in Locke's *Second Treatise of Government*. Locke's treatise, which was published in 1690, is one of the classic texts of constitutional law. Here we see that the man who became, in Keynes's words, "Sage and Monarch of the Age of Reason...the eighteenth-century Sir Isaac, so remote from the child magician born in the first half of the seventeenth century," was also one of the architects of our civil liberties. And for Newton, the struggle for political freedom was never separated from the struggle for a true understanding of God.

The best and most original part of Gleick's book is the description of the young Newton in his first five chapters. Gleick's account is based on a detailed study of the manuscript notebooks that Newton kept as a student in Cambridge, recording his many false starts and digressions as he groped his way toward an understanding of the laws of nature. In these notebooks we see him, not yet possessing words to express the concepts such as force and momentum that would allow him to formulate the laws precisely, and not yet possessing the mathematical tools of differential and integral calculus that would allow him to deduce the consequences of the laws. To reach his fundamental insight that the laws of nature can be expressed as differential equations, he had to simultaneously guess the laws and invent the mathematical language of calculus in which to express them. The notebooks record his successes and failures as they happened, not reinterpreted in the light of later discoveries.

It is lucky for us that Newton was working alone, without friends or collaborators, sharing his intellectual adventures with nobody. Instead of telling his thoughts to friends, he told them to his notebooks. In the notebooks we see the slow dawning of his understanding, and then the rapid succession of discoveries leading up to the breakthrough of 1665 and 1666, the plague years, when Newton left Cambridge to escape the plague and stayed at his home in Woolsthorpe. At Woolsthorpe, at the age of twenty-four, he put together the pieces and assembled his new vision of the universe. The story of these five years, from Newton's arrival at Cambridge as a student in 1661 to his solitary triumph at Woolsthorpe in 1666, is told more clearly by Gleick than by Westfall. Gleick has gone back to the original notebooks and brought them to life.

In 1667 Newton became a fellow of Trinity College and resumed his solitary existence in Cambridge. He bought apparatus and materials for the alchemical experiments that occupied much of his time for the next twenty years. He spoke to nobody about his alchemical studies, and to almost nobody about his discoveries in physics. For him, alchemy and physics and theology were parts of a single enterprise, three aspects of a single search for knowledge that God had placed within his grasp. Since he was not free to talk about his theology, he saw no reason why he should talk about his alchemy or his physics. He might never have talked about his physics, if his friend Halley had not come to Cambridge in 1684 begging him to publish what he knew. Then, once he had started writing down his physical discoveries in logical sequence, he did not stop until he had finished the three volumes of the *Principia*.

At the beginning of the seventeenth century, the birth of modern science had been proclaimed by two great philosophers, Francis Bacon in England and René Descartes in France. Bacon and Descartes had very different visions of how science should be pursued. According to Bacon, scientists should experiment freely and collect facts about everything in the world, until in due time the accumulation of facts would make clear the way nature behaves. From the storehouse of accumulated facts, scientists would induce the laws of nature. According to Descartes, scientists should deduce the laws of nature by pure reason, starting from the axioms of mathematics and our knowledge of the existence of God. Experiments needed to be done only to verify that the logical deduction of the laws of nature was correct. During the seventeenth century, science in England tended to follow the Baconian path, with the Royal Society in London collecting facts about everything from two-headed calves to rainstorms of frogs and fish. Science in France followed the Cartesian path, and was dominated by Descartes's theory of vortices. The Cartesian vortices were supposed to fill space on Earth and in the heavens, pushing celestial objects along their orbits in the sky. At the time when Newton made his discoveries, the learned men of England were mostly doing science in the empirical style of Bacon, but most of them believed in the Cartesian theory of vortices because it was the only theory available.

Newton himself was at heart a Cartesian, reaching his insights into the nature of things by pure thought as Descartes intended. When he came to write the *Principia*, he wrote it in Cartesian style, stating his conclusions in the form of propositions and theorems, and using the methods of pure geometry to prove them. But unlike Descartes, he was himself an experimenter and understood the importance of precise experiments for testing theories. So, in the *Principia*, he brilliantly succeeded in using the Cartesian method to demolish the Cartesian theory. In the first two volumes he built a grand edifice of mathematics, more coherent than anything Descartes had to offer, and then in the third volume he delivered the coup de grâce, demonstrating with an abundance of observational facts that nature danced to his tune. As soon as the *Principia* was published and widely circulated, the Cartesian vortices were dead.

Newton was a skillful fighter and always played to win. He enjoyed his victories over Descartes and King James. He also enjoyed victories over Robert Hooke, who claimed to have anticipated him in the discovery of the law of universal gravitation, and over Gottfried Leibniz, who claimed to have anticipated him in the discovery of calculus. As Master of the Mint, he zealously prosecuted counterfeiters of the coinage, rejected their pleas for clemency, and made sure they were hanged. He went out of his way not only to defeat his opponents but to crush and humiliate them. I imagine him now, wherever he may be in the spiritual realms of heaven or hell, enjoying his final victory over Lord Lymington. Lord Lymington attempted to profit at Newton's expense, scattering his papers to the winds for a paltry £9,000. The final result of Lord Lymington's impiety is that he is remembered as a Judas who betrayed his master, while Newton's papers are preserved and studied by a multitude of scholars as never before.

Postscript, 2006

In response to this review, I received a number of informative letters from Newton scholars who know more about Newton than I do. I have corrected the review where they found mistakes. Robert Iliffe, director of the Newton Manuscript Project at Imperial College in London, informs me that the Babson papers are now on a semi-permanent loan to the Dibner Institute at MIT, where they are conveniently accessible. I am grateful to Sarah Jones Nelson for showing me the manuscript "Magdalen MW 432" which she discovered in the Magdalen College archives in Oxford. She published a brief description of the manuscript in the *Magdalen College Record*, 2001, pages 102–104.

18

CLOCKWORK SCIENCE

TODAY THE NAME of Albert Einstein is known to almost everybody, the name of Henri Poincaré to almost nobody. A hundred years ago the opposite was true. Then, Einstein was a newly appointed technical expert, third class, examining patent applications in the Swiss patent office in Bern, having failed in his efforts to find an academic job, while Poincaré was one of the leading figures of the French scientific establishment, famous not only as a great scientist but as the author of popular books that were translated into many languages and kept the public informed about the dramatic progress of science during the early years of the twentieth century. In 1903, Einstein and Poincaré were both working hard at one of the central problems of science, trying to find a correct theory to describe how fast particles behave in electric and magnetic fields. Poincaré had published several papers on the subject which Einstein may or may not have read. Einstein had published nothing.

Two years later, in 1905, Poincaré and Einstein simultaneously arrived at a solution to the problem. Poincaré presented a summary of his results to the French Academy of Sciences in Paris, and in the same month Einstein mailed his classic paper, "Electrodynamics of Moving Bodies," to the German journal *Annalen der Physik*. The two versions of the solution were in substance almost identical. Both were based on the principle of relativity, which says that the laws of nature are the same for a moving observer as they are for an observer standing still. Both agreed with the experimentally observed behavior of fast particles, and made the same predictions for the results of future experiments. How then did it happen that Einstein became world famous as the discoverer of relativity, while Poincaré did not? Poincaré's lasting fame, such as it is, derives from his discoveries in other areas of science and not from his work on relativity. Is the verdict of posterity, giving all the credit for relativity to Einstein and none to Poincaré, fair or unfair? I will return to these questions later.

Peter Galison is a historian and not a judge. His purpose is to understand the way in which Poincaré and Einstein arrived at their insights, not to hand out praise or blame. His *Einstein's Clocks, Poincaré's Maps: Empires of Time*[1] is an extended double portrait, describing their lives and times in detail. At the beginning, he complains of the unequal treatment given to them by biographers: "There are, to be sure, too many biographies of Einstein and not enough of Poincaré." Poincaré was a great man who lived a full and many-sided life, and he deserves at least a fraction of the attention that has been lavished on Einstein. For readers who may be interested in learning more about Poincaré, I recommend a short biography by Benjamin Yandell that Galison does not mention. Yandell's book, *The Honors Class: Hilbert's Problems and Their Solvers*,[2] is a collection of biographies of mathematicians who solved a famous list of twenty-three problems propounded by David Hilbert at the International Congress of Mathematicians in Paris in 1900. Poincaré solved problem number 22. The Poincaré biography in Yandell's book is one of the best. It gives us in thirty pages a vivid picture of Poincaré's life as a mathematician, and overlaps very little with Galison's account.

---

1. Norton, 2003.

2. A.K. Peters, 2002.

Galison's book does not contain a single equation. Anybody with an interest in history can read it. The stories that it tells are mostly about the applications of science and not about science itself. The applications are things that everybody can understand, maps in the case of Poincaré, clocks in the case of Einstein. Poincaré and Einstein were both deeply involved in the practical world of electrical communication and machines at the same time as they were working out the theory of relativity. The scientific theories that emerged from their practical concerns are described briefly, without any mathematical or technical jargon. The book consists of six chapters, four long chapters in the middle and two short chapters at the beginning and the end. The two short chapters set the stage and summarize the conclusions. The little that is said about the details of the theory of relativity is mostly contained in the two short chapters. The main conclusion is that, in the events that led to the discovery of the theory, philosophical speculation and technological invention were inextricably mixed.

The four long chapters consist of a collection of stories, describing how electricity transformed the world in the second half of the nineteenth century. Here is a typical story. A political battle was fought in Hartford, Connecticut, in 1880, to decide whether the trains in Connecticut should run on Boston time or New York time. The battle was fought between the Harvard astronomer Leonard Waldo and the New York to Hartford railroad. Waldo was not only an astronomer but also an entrepreneur. His observatory ran a business, selling accurate time signals that were distributed to customers by electric telegraph. The customers were railroads and city fire departments, manufacturers and retailers of clocks and watches, and private citizens who possessed fine watches and wished to check their accuracy. Waldo made a big pitch to the Hartford city council, emphasizing the superior precision of Boston time. But the New York to Hartford railroad ran on New York time and would not be moved. The railroad won the battle.

Another story tells how the United States led the world in adopting a unified system of time zones in 1883, with the times in neighboring zones differing by exactly one hour. This was another victory for the railroads. The crucial convention to decide the fate of the time-zone system was held in St. Louis, Missouri. The voting was measured not by the number of delegates voting but by the miles of railroad track that they represented. The final vote was 79,041 miles of track in favor and 1,714 miles opposed. After that, cities were compelled to set their clocks by railroad time and not by local time. Even New York City gave up its local time and set its clocks to astronomical time at longitude seventy-five west.

A third story tells how the French Bureau of Longitude established a commission in the year 1897 to extend the metric system to the measurement of time. The idea was to abolish the antiquated division of the day into hours, minutes, and seconds, and replace it by a division into tenths, thousandths, and hundred thousandths of a day. The new second would be a hundred thousandth of a day, and the new hour would be ten kiloseconds. The new units of time could then be handled as conveniently as grams and kilograms, meters and kilometers. To convert time from days to hours or minutes or seconds, we would only need to move a decimal point. We would no longer need to struggle with multiplications and divisions by twenty-four and sixty. This was a revival of a dream that was in the minds of the creators of the metric system at the time of the French Revolution a hundred years earlier. Some members of the Bureau of Longitude commission introduced a compromise proposal, retaining the old-fashioned hour as the basic unit of time and dividing it into hundredths and ten thousandths. Poincaré served as secretary of the commission and took its work very seriously, writing several of its reports. He was a fervent believer in a universal metric system. But he lost the battle. The rest of the world outside France gave no support to the commission's proposals, and the French government was not prepared to go it alone. After three years of hard work, the commission was dissolved in 1900.

These stories, and many others, are told to illustrate Galison's thesis that the coordination of time signals was a central concern of people and governments in the later part of the nineteenth century. So it was no accident that the coordination of time signals also has a central role in the theory of relativity. Poincaré and Einstein lived in a period of history when the transmission of time signals was a growth industry, and both of them were professionally involved in it. Poincaré worked for the Bureau of Longitude, which was responsible for the mapping of French territories around the globe. To make accurate maps, the bureau needed to determine accurate longitudes. To determine longitude at a remote place such as Dakar or Haiphong, it was necessary to compare the local time, obtained from local astronomical measurements, with Paris time, obtained by receiving an accurate time signal from Paris. The accuracy of maps therefore depended on the accuracy of long-distance transmission of time signals. The transmission of time signals, first by overland telegraph lines and undersea cables, and later by radio, was a difficult technical problem. Signals were attenuated by transmission losses and corrupted by ambient noise. The transmission introduced delays which had to be accurately calculated, so that Paris time could be correctly deduced from the observed time of reception of signals. The recording apparatus introduced other delays which had to be measured and compensated. The transmission of time signals with high accuracy required a mastery of both theory and practical engineering.

Poincaré was well versed in practice as well as theory. He started his professional career as a mining engineer, inspecting mines in the coal fields of northern France. One of his first jobs was to investigate a disastrous explosion that killed eighteen miners. He descended into the mine looking for clues while the corpses were still warm. He found a miner's lamp which had a rectangular hole in its wire mesh, apparently caused by a blow from a pickax. The wire mesh, a device invented by Humphrey Davy sixty-five years earlier, prevents the flame inside the lamp from igniting explosive gases in the mine outside the lamp. The mesh allows air to pass through but stops flame from propagating from inside to outside. When the mesh was broken, the flame could propagate freely and all hell broke loose. Poincaré risked his life to discover what had happened, and wrote a report analyzing in detail the flow of gases in the mine.

Einstein grew up in a family of electrical engineers. His father and uncle ran a business in Munich, manufacturing and selling electrical measuring equipment. One of his uncle Jakob's patents dealt with equipment for the electrical control of clocks. Einstein's early familiarity with electrical machinery helped him to get his job at the Swiss patent office, and helped him to do the job well. As soon as he started work, he was confronted with numerous applications for patents concerned with electric clocks and with their coordination by distribution of electric time signals. In the year 1904, when the theory of relativity was in process of gestation, fourteen such patents were approved by the Bern office. The number of applications that were disapproved is not recorded.

At that time, Switzerland was becoming a world leader in the manufacture of precision clocks, and applications for Swiss patents were pouring in from hopeful inventors all over the world. For Einstein, analyzing and understanding these inventions was not just a convenient way to pay the rent. He enjoyed the work at the patent office and found it intellectually challenging. Later in his life, he remarked that the formulation of technological patents had been an important stimulus to his thinking about physics.

Among historians of science during the last half-century, there have been two predominant schools of thought. The leaders of the two schools were Thomas Kuhn and Peter Galison. Kuhn, in his classic work *The Structure of Scientific Revolutions*, published in 1962, portrays the progress of science as a kind of punctuated equilibrium, like the evolution of species in the history of life. Most of the time, evolution is slow or stagnant, the species are well adapted to their environments, and natural selection keeps them from changing fast. Then, when the environment is disturbed and new ecological niches are opened, selection favors rapid change, and small populations of lucky individuals change rapidly enough to form new species. So in science, the normal state of affairs is a slowly changing equilibrium, with a dominant orthodox theory that explains observed phenomena and is not seriously questioned. So long as normal science prevails, the job of the scientist is to solve unimportant puzzles that arise within the accepted dogma. But at rare moments, new discoveries or new ideas arise that call the accepted dogma into question, and then a scientific revolution may occur. To cause a scientific revolution, the new discoveries must be powerful enough to overthrow the prevailing theory, and a new set of ideas must be ready to replace it. In Kuhn's view, it is new ideas that drive scientific revolutions. The big steps forward in the progress of science are idea-driven.

In contrast to Kuhn, Galison in his classic work *Image and Logic*, published in 1997, describes the history of particle physics as a history of tools rather than ideas. According to *Image and Logic*, the progress of science is tool-driven. The tools of particle physics are of two kinds, optical and electronic. The optical tools are devices such as cloud chambers, bubble chambers, and photographic emulsions, which display particle interactions visually by means of images. The images record the tracks of particles. An experienced experimenter can see at once from the image when a particle is doing something unexpected. Optical tools are more likely to lead to discoveries that are qualitatively new.

On the other hand, electronic tools are better for answering quantitative questions. Electronic detectors such as the Geiger counters that measure radioactivity in the cellars of old houses are based on logic. They are programmed to ask simple questions each time they detect a particle, and to record whether the answers to the questions are yes or no. They can detect particle collisions at rates of millions per second, sort them into yes's and no's, and count the number that answered yes and the number that answered no. The history of particle physics may be divided into two periods, the earlier period ending about 1980 when optical detectors and images were dominant, and the later period when electronic detectors and logic were dominant. Before the transition, science advanced by making qualitative discoveries of new particles and new relationships between particles. After the transition, with the zoo of known particles more or less complete, the science advanced by measuring their interactions with greater and greater precision. In both periods, before and after the transition, tools were the driving force of progress.

The arguments between Kuhnian historians emphasizing ideas and Galisonian historians emphasizing tools have continued to be vigorous. Historians trained in theoretical science tend to be Kuhnians, while those trained in experimental science tend to be Galisonians. Whether one chooses to emphasize ideas or tools is to some extent a matter of taste. I myself tend to be a Galisonian, although I am a theoretician by training. But in this debate, as often happens when academic scholars engage in disputes, the disciples of each leader are more dogmatic than the leaders. I once attended a meeting of historians at which the disciples of Kuhn were presenting an extreme and exaggerated version of his views. Kuhn interrupted them by shouting from the back of the hall with overwhelming volume, "One thing you people need to understand: I am not a Kuhnian."

Kuhn believed in the primacy of ideas, but not to the exclusion of everything else. And in his new book, Galison is telling us that he still believes in the primacy of tools, but not to the exclusion of everything else. As I came to the final chapter of the book, I could almost hear him shouting, "One thing you people need to understand: I am not a Galisonian."

Galison uses the phrase "critical opalescence" to sum up the story of what happened in 1905 when relativity was discovered. Critical opalescence is a strikingly beautiful effect that is seen when water is heated to a temperature of 374 degrees Celsius under high pressure. 374 degrees is called the critical temperature of water. It is the temperature at which water turns continuously into steam without boiling. At the critical temperature and pressure, water and steam are indistinguishable. They are a single fluid, unable to make up its mind whether to be a gas or a liquid. In that critical state, the fluid is continually fluctuating between gas and liquid, and the fluctuations are seen visually as a multicolored sparkling. The sparkling is called opalescence because it is also seen in opal jewels which have a similar multicolored radiance.

Galison uses critical opalescence as a metaphor for the merging of technology, science, and philosophy that happened in the minds of Poincaré and Einstein in the spring of 1905. Poincaré and Einstein were immersed in the technical tools of time signaling, but the tools by themselves did not lead them to their discoveries. They were immersed in the mathematical ideas of electrodynamics, but the ideas by themselves did not lead them to their discoveries. They were also immersed in the philosophy of space and time. Poincaré had written a philosophical book, *Science and Hypothesis*, which Einstein studied, digging deep into the foundations of knowledge and criticizing the Newtonian notions of absolute space and time. But the philosophy by itself did not lead them to their discoveries. What was needed to give birth to the theory of relativity was a critical moment, when tools, ideas, and philosophical reflections jostled together and merged into a new way of thinking. Galison would like to put an end to the argument between Kuhnians and Galisonians. In this book he takes his position squarely in the middle: "Attending to moments of critical opalescence offers a way out of this endless oscillation between thinking of history as ultimately about ideas or fundamentally about material objects."

The one question that Galison's metaphor of critical opalescence does not answer is why Einstein discovered the theory of relativity as we know it and Poincaré did not. The theories discovered by Poincaré and Einstein were operationally equivalent, with identical experimental consequences, but there was one crucial difference. The difference was the use of the word "ether." The wave theory of light, and the theories of electric and magnetic forces that were developed in the nineteenth century, were all based on the idea of ether. James Clerk Maxwell, who unified the theories of light and electromagnetism in 1865, was a firm believer in ether. Electric and magnetic forces behaved like mechanical stresses in a solid medium with suitable properties of rigidity and elasticity. Therefore, it was believed, a solid medium must exist, pervading the whole of space and carrying the electric and magnetic stresses. Light waves must be shear waves in the same elastic medium. The all-pervading solid medium was given the name "ether."

The essential difference between Poincaré and Einstein was that Poincaré was by temperament conservative and Einstein was by temperament revolutionary. When Poincaré looked for a new theory of electromagnetism, he tried to preserve as much as he could of the old. He loved the ether and continued to believe in it, even when his own theory showed that it was unobservable. His version of relativity theory was a patchwork quilt. The new idea of local time, depending on the motion of the observer, was patched onto the old framework of absolute space and time defined by a rigid and immovable ether. Einstein, on the other hand, saw the old framework as cumbersome and unnecessary and was delighted to be rid of it. His version of the theory was simpler and more elegant. There was no absolute space and time and there was no ether. All the complicated explanations of electric and magnetic forces as elastic stresses in the ether could be swept into the dustbin of history, together with the famous old professors who still believed in them. All local times were equally valid. In order to calculate with Einstein's version of relativity, all you needed to know was the rule for transforming from one local time to another. In the competition for public recognition, the clarity and simplicity of Einstein's argument gave him an overwhelming advantage.

Poincaré and Einstein only met once, at a conference in Brussels in 1911. The meeting did not go well. Einstein afterward reported his impression of Poincaré: "Poincaré was simply negative in general, and, all his acumen notwithstanding, he showed little grasp of the situation." So far as Einstein was concerned, Poincaré belonged with the ether in the dustbin of history. But Einstein underestimated Poincaré. Einstein did not know that Poincaré had just then written a letter recommending him for a professorship at the Swiss Federal Institute of Technology in Zürich. Here is what Poincaré had to say about Einstein:

> What we must above all admire in him, is the facility with which he has adapted to new conceptions and from which he knows how to draw the consequences. He does not remain attached to classical principles, and, in the presence of a problem of physics, is prompt to envision all the possibilities.... The future will show more and more the value of Mr. Einstein, and the university that finds a way to secure this young master is assured of drawing from it great honor.

Poincaré bore no grudge against his young rival. He was still driven by the same generous impulse that made him rush into the coal mine at Magny thirty-two years earlier. A year after the meeting with Einstein in Brussels, Poincaré was dead. Einstein never saw Poincaré's letter and never knew that he had misjudged him.

Looking back upon this history, I disagree with Galison's conclusion. I do not see critical opalescence as a decisive factor in Einstein's victory. I see Poincaré and Einstein equal in their grasp of contemporary technology, equal in their love of philosophical speculation, unequal only in their receptiveness to new ideas. Ideas were the decisive factor. Einstein made the big jump into the world of relativity because he was eager to throw out old ideas and bring in new ones. Poincaré hesitated on the brink and never made the big jump. In this instance at least, Kuhn was right. The scientific revolution of 1905 was driven by ideas and not by tools.[3]

3. The theme of this review, the question whether tools or ideas were dominant in the revolution of 1905, is discussed in a wider context in the chapter "Scientific Revolutions" in my book *The Sun, the Genome and the Internet* (Oxford University Press, 1999). There I came to the conclusion that the majority of revolutions are tool-driven, the revolution of 1905 being one of the notable exceptions.

19

THE WORLD ON A STRING

IN THE GOLDEN years of the Liberal Party in England, before the First World War, Herbert Asquith was the patrician prime minister and Winston Churchill was an obstreperous young politician. At question time in the House of Commons, Churchill frequently challenged Asquith with provocative statements and awkward questions. After one of these Churchillian assaults, Asquith lamented, "I wish I knew as much about anything as that young man knows about everything." Reading *The Fabric of the Cosmos: Space, Time, and the Texture of Reality*,[1] this eloquent book in which Brian Greene lays out before us his vision of the cosmos, I feel some sympathy for Asquith. Asquith expresses my reaction to the book precisely.

I recommend Greene's book to any nonexpert reader who wants an up-to-date account of theoretical physics, written in colloquial language that anyone can understand. For the nonexpert reader, my doubts and hesitations are unimportant. It is not important whether Greene's picture of the universe will turn out to be technically accurate. The important thing is that his picture is coherent and intelligible and consistent with recent observations. Even if many of the details later turn out to be wrong, the picture is a big step toward

1. Knopf, 2004.

understanding. Progress in science is often built on wrong theories that are later corrected. It is better to be wrong than to be vague. Greene's book explains to the nonexpert reader two essential themes of modern science. First it describes the historical path of observation and theory that led from Newton and Galileo in the seventeenth century to Einstein and Stephen Hawking in the twentieth. Then it shows us the style of thinking that led beyond Einstein and Hawking to the fashionable theories of today. The history and the style of thinking are authentic, whether or not the fashionable theories are here to stay.

In his book *The Elegant Universe*, published in 1999, Greene gave us a more detailed and technical account of string theory, the theory to which his professional life as a physicist has been devoted. The earlier book was remarkably successful in translating the abstruse and abstract ideas of string theory into readable prose. Early in his new book he gives a brief summary of string theory as he expounded it in *The Elegant Universe*:

> Superstring theory starts off by proposing a new answer to an old question: what are the smallest, indivisible constituents of matter? For many decades, the conventional answer has been that matter is composed of particles—electrons and quarks—that can be modeled as dots that are indivisible and that have no size and no internal structure. Conventional theory claims, and experiments confirm, that these particles combine in various ways to produce protons, neutrons, and the wide variety of atoms and molecules making up everything we've ever encountered.
>
> Superstring theory tells a different story. It does not deny the key role played by electrons, quarks, and the other particle species revealed by experiment, but it does claim that these particles are not dots. Instead, according to superstring theory, every particle is composed of a tiny filament of energy, some hundred billion billion times smaller than a single atomic nucleus (much smaller

> than we can currently probe), which is shaped like a little string. And just as a violin string can vibrate in different patterns, each of which produces a different musical tone, the filaments of superstring theory can also vibrate in different patterns. But these vibrations don't produce different musical notes; remarkably, the theory claims that they produce different particle properties. A tiny string vibrating in one pattern would have the mass and the electric charge of an electron; according to the theory, such a vibrating string would be what we have traditionally called an electron. A tiny string vibrating in a different pattern would have the requisite properties to identify it as a quark, a neutrino, or any other kind of particle. All species of particles are unified in superstring theory since each arises from a different vibrational pattern executed by the same underlying entity.

This is a fine beginning for a theory of the universe, and maybe it is true. To be useful, a scientific theory does not need to be true, but it needs to be testable. My doubts about string theory arise from the fact that it is not at present testable. Greene discusses in his Chapters 13 and 14 the prospects for experimental tests of the theory. The experiments that he describes will certainly open new doors to the understanding of nature, even if they do not answer the question whether string theory is true.

The Fabric of the Cosmos covers a wider field than *The Elegant Universe* and paints it with a broader brush. There is not much overlap between the two books. Only Chapter 12 of the new book, which summarizes the earlier book and gives us the gist of string theory without the details, overlaps strongly. Greene himself suggests that readers who have read *The Elegant Universe* should skim through Chapter 12. Except for this chapter, the two books cover different subjects and can be read independently. Neither is a prerequisite for reading the other. The new book is easier, and should preferably be read first. Readers who got stuck halfway through *The Elegant Universe* may find the new book more digestible.

In the history of science there is always a tension between revolutionaries and conservatives, between those who build grand castles in the air and those who prefer to lay one brick at a time on solid ground. The normal state of tension is between young revolutionaries and old conservatives. This is the way it is now, and the way it was eighty years ago when the quantum revolution happened. I am a typical old conservative, out of touch with the new ideas and surrounded by young string theorists whose conversation I do not pretend to understand. In the 1920s, the golden age of quantum theory, the young revolutionaries were Werner Heisenberg and Paul Dirac, making their great discoveries at the age of twenty-five, and the old conservative was Ernest Rutherford, dismissing them with his famous statement, "They play games with their symbols but we turn out the real facts of Nature." Rutherford was a great scientist, left behind by the revolution that he had helped to bring about. That is the normal state of affairs.

Fifty years ago, when I was considerably younger than Greene is now, things were different. The normal state of affairs was inverted. At that time, in the late 1940s and early 1950s, the revolutionaries were old and the conservatives were young. The old revolutionaries were Albert Einstein, Dirac, Heisenberg, Max Born, and Erwin Schrödinger. Every one of them had a crazy theory that he thought would be the key to understanding everything. Einstein had his unified field theory, Heisenberg had his fundamental length theory, Born had a new version of quantum theory that he called reciprocity, Schrödinger had a new version of Einstein's unified field theory that he called the Final Affine Field Laws, and Dirac had a weird version of quantum theory in which every state had probability either plus two or minus two. Probability, as common sense defines it, is a number between zero and one expressing our degree of confidence that an event will happen. Probability one means that the event always happens; probability zero means that it never happens. In Dirac's Alice-in-Wonderland world, every state happens either more often than always or less often than never. Each of the five old men believed that physics needed another revolution as profound as the quantum revolution that they had led twenty-five years earlier. Each of them believed that his pet idea was the crucial first step along a road that would lead to the next big breakthrough.

Young people like me saw all these famous old men making fools of themselves, and so we became conservatives. The chief young players then were Julian Schwinger and Richard Feynman in America and Sin-Itiro Tomonaga in Japan. Anyone who knew Feynman might be surprised to hear him labeled a conservative, but the label is accurate. Feynman's style was ebullient and wonderfully original, but the substance of his science was conservative. He and Schwinger and Tomonaga understood that the physics they had inherited from the quantum revolution was pretty good. The physical ideas were basically correct. They did not need to start another revolution. They only needed to take the existing physical theories and clean up the details. I helped them with the later stages of the cleanup. The result of our efforts was the modern theory of quantum electrodynamics, the theory that accurately describes the way atoms and radiation behave.

This theory was a triumph of conservatism. We took the theories that Dirac and Heisenberg had invented in the 1920s, and changed as little as possible to make the theories self-consistent and user-friendly. Nature smiled on our efforts. When new experiments were done to test the theory, the results agreed with the theory to eleven decimal places. But the old revolutionaries were still not convinced. After the results of the first experiments had been announced, I brashly accosted Dirac and asked him whether he was happy with the big success of the theory that he had created twenty-five years earlier.

Dirac, as usual, stayed silent for a while before replying. "I might have thought that the new ideas were correct," he said, "if they had not been so ugly." That was the end of the conversation. Einstein too was unimpressed by our success. During the time that the young physicists at the Institute for Advanced Study in Princeton were deeply engaged in developing the new electrodynamics, Einstein was working in the same building and walking every day past our windows on his way to and from the institute. He never came to our seminars and never asked us about our work. To the end of his life, he remained faithful to his unified field theory.

Looking back on this history, I feel no shame in being a conservative today. I belong to a generation that saw conservatism triumph, and I remain faithful to our ideals just as Einstein remained faithful to his. But now my generation is passing from the scene, and I am wondering what the next cycle of history will bring. After the revolutionaries of string theory have grown old, what will the next generation think of them? Will there be another generation of young revolutionaries? Or shall we again have an inversion of the normal state of things, with a new generation of young conservatives in rebellion against the elderly pioneers of string theory? My generation will not be around to see these questions answered.

One of the main themes in Greene's book is the disconnect between Einstein's theory of general relativity and quantum mechanics, the two discoveries that revolutionized physics at the beginning of the twentieth century. Einstein's theory is primarily a theory of gravity, describing the gravitational field as a curvature of space-time, and describing the fall of an apple as the response of the apple to the curvature of space-time induced by the mass of the earth. Einstein's theory treats the apple and the earth as classical objects with precisely defined positions and velocities, paying no attention to the uncertainties introduced by quantum mechanics. The apple and the earth are large enough so that the quantum uncertainties are negligible.

On the other hand, quantum mechanics describes the behavior of atoms and elementary particles, for which the quantum uncertainties have a dominating influence, and pays no attention to gravity. The atoms and particles are small enough so that any gravitational fields that they induce are negligible. The two theories divide the universe of physics between them without overlapping, general relativity taking care of large objects from apples to galaxies, and quantum mechanics taking care of small objects from molecules to light-quanta. General relativity is important for astronomy and cosmology, while quantum mechanics is important for atomic physics and chemistry. This division of the universe works well for all practical purposes. It works well because the gravitational effects of single atoms or particles are unobservably small.

Greene takes it for granted, and here the great majority of physicists agree with him, that the division of physics into separate theories for large and small objects is unacceptable. General relativity is based on the idea that space-time is a flexible structure pulled and pushed by material objects. Quantum mechanics is based on the idea that space-time is a rigid framework within which observations are made. The two theories are mathematically incompatible. Greene believes that there is an urgent need to find a theory of quantum gravity that applies to large and small objects alike. Quantum gravity means a unified theory that works like general relativity for large objects and like quantum mechanics for small objects. In spite of heroic efforts by many people, no consistent theory of quantum gravity was found until string theory came along. The first and greatest triumph of string theory was its success in unifying general relativity with quantum mechanics. That success gave its discoverers some justification for claiming that it could be a "theory of everything." String theory is still incomplete and far from ready for practical application, but it does in principle provide us with a theory of quantum gravity.

As a conservative, I do not agree that a division of physics into separate theories for large and small is unacceptable. I am happy with the situation in which we have lived for the last eighty years, with separate theories for the classical world of stars and planets and the quantum world of atoms and electrons. Instead of insisting dogmatically on unification, I prefer to ask the question whether a unified theory would have any real physical meaning. The essence of any theory of quantum gravity is that there exists a particle called the graviton which is a quantum of gravity, just like the photon which is a quantum of light. Such a particle is necessary in quantum gravity, because energy is carried in discrete little packets called quanta, and a quantum of gravitational energy would behave like a particle.

The question that I am asking is whether there is any conceivable way in which we could detect the existence of individual gravitons. It is easy to detect individual photons, as Einstein showed, by observing the behavior of electrons kicked out of metal surfaces by light incident on the metal. The difference between photons and gravitons is that gravitational interactions are enormously weaker than electromagnetic interactions. If you try to detect individual gravitons by observing electrons kicked out of a metal surface by incident gravitational waves, you find that you have to wait longer than the age of the universe before you are likely to see a graviton. I looked at various possible ways of detecting gravitons and did not find a single one that worked. Because of the extreme weakness of the gravitational interaction, any putative detector of gravitons has to be extravagantly massive. If the detector has normal density, most of it is too far from the source of gravitons to be effective, and if it is compressed to a high density around the source it collapses into a black hole. There seems to be a conspiracy of nature to prevent the detector from working.

I propose as a hypothesis to be tested that it is impossible in principle to observe the existence of individual gravitons. I do not claim that this hypothesis is true, only that I can find no evidence against it. If it is true, quantum gravity is physically meaningless. If individual gravitons cannot be observed in any conceivable experiment, then they have no physical reality and we might as well consider them nonexistent. They are like the ether, the elastic solid medium which nineteenth-century physicists imagined filling space. Electric and magnetic fields were supposed to be tensions in the ether, and light was supposed to be a vibration of the ether. Einstein built his theory of relativity without the ether, and showed that the ether would be unobservable if it existed. He was happy to get rid of the ether, and I feel the same way about gravitons.

According to my hypothesis, the gravitational field described by Einstein's theory of general relativity is a purely classical field without any quantum behavior. Gravitational waves exist and can be detected, but they are classical waves and not collections of gravitons. If this hypothesis is true, we have two separate worlds, the classical world of gravitation and the quantum world of atoms, described by separate theories. The two theories are mathematically different and cannot be applied simultaneously. But no inconsistency can arise from using both theories, because any differences between their predictions are physically undetectable.

Another major theme of Greene's book is the interpretation of quantum mechanics and the weird phenomena of quantum entanglement. He devotes two long chapters, "Entangling Space" and "Time and the Quantum," to this theme. He makes a valiant attempt to clarify a notoriously foggy subject. But he makes his task more difficult by insisting that quantum mechanics must include everything. He rejects without any serious discussion the dualistic interpretation of quantum mechanics, the idea that there are two separate worlds, the classical world and the quantum world, each following its own rules. The dualistic view, limiting the scope of quantum mechanics to well-defined experimental situations, makes the problems of interpretation much simpler.

The dualistic interpretation of quantum mechanics says that the classical world is a world of facts while the quantum world is a world of probabilities. Quantum mechanics predicts what is likely to happen while classical mechanics records what did happen. This division of the world was invented by Niels Bohr, the great contemporary of Einstein who presided over the birth of quantum mechanics. Lawrence Bragg, another great contemporary, expressed Bohr's idea more simply: "Everything in the future is a wave, everything in the past is a particle." Since the greater part of our knowledge is knowledge of the past, Bohr's division limits the scope of quantum mechanics to a small part of science. I like Bohr's division, because it allows the possibility that gravitons may not exist. If the scope of quantum theory is limited, gravity may legitimately be excluded from it. But Greene will not accept any such limitation. After briefly describing Bohr's point of view, he says:

> For decades, this perspective held sway. However, its calmative effect on the mind struggling with quantum theory notwithstanding, one can't help feeling that the fantastic predictive power of quantum mechanics means that it is tapping into a hidden reality that underlies the workings of the universe.

I prefer the calmative effect of Bohr's perspective on the mind, while Greene prefers the hidden reality. In his first chapter, Greene shows us what he means by hidden reality:

> Superstring theory combines general relativity and quantum mechanics into a single, consistent theory.... And as if that weren't enough, superstring theory has revealed the breadth necessary to stitch all of nature's forces and all of matter into the same theoretical tapestry. In short, superstring theory is a prime candidate for Einstein's unified theory.
>
> These are grand claims, and, if correct, represent a monumental step forward. But the most stunning feature of superstring

> theory, one that I have little doubt would have set Einstein's heart aflutter, is its profound impact on our understanding of the fabric of the cosmos.... Instead of the three spatial dimensions and one time dimension of common experience, superstring theory requires nine spatial dimensions and one time dimension.... As we don't see these extra dimensions, superstring theory is telling us that we've so far glimpsed but a meager slice of reality.

The next-to-last chapter, "Teleporters and Time Machines," is a pleasant interlude, describing some possible engineering applications of quantum entanglement and general relativity. The teleporter is a device that can scan an object at one place and reproduce a precise copy of it at another place far away, using quantum entanglement to ensure that the reproduction is exact. The good news is that such a device is in principle possible. The bad news is that it inevitably destroys the object that it copies. The time machine is a tunnel through hyperspace connecting two portals that exist at different places and times in our universe. If you can find the portal that is later in time, you can walk through the tunnel to emerge in your own past. The good news is that such a tunnel is a possible solution of the equations of general relativity. The bad news is that a tunnel large enough to walk through would require more than the total energy output of the sun to hold it open. Neither the teleporter nor the time machine is likely to contribute much to the welfare of our descendants. Greene describes these fantasies with a proper mixture of scientific accuracy and irony.

In January 2001, I was invited to the World Economic Forum in Davos, Switzerland. Brian Greene was also invited, and we were asked to hold a public debate on the question "When will we know it all?" In other words, when will the last big problems of science be solved? The audience consisted mainly of industrial and political tycoons. Our debate was intended to entertain the tycoons, not to give them a serious scientific education. To make it more amusing, Greene was asked to take an extreme position saying "Soon," and I was asked to take an extreme position saying "Never."

Here is my version of Greene's opening statement, reconstructed from my unreliable memory after we came back from Switzerland. He said that this generation of scientists is amazingly lucky. Within a few years or decades, we will discover the fundamental laws of nature. The fundamental laws will be a finite set of equations, like Maxwell's equations of electrodynamics or Einstein's equations of gravitation. Everything else will then follow from these equations. Once we have the fundamental equations, we are done. If we are not smart enough to find the equations, then we will leave it to our grandchildren to finish the job. Either way, the end of fundamental science is near.

Greene said his confidence in our ability to find the fundamental laws is based on the marvelous fact that the laws of nature are simple and beautiful. The history of physics shows that this is true of all the laws that we have discovered in the past. We did not need to do unending experiments to discover the laws. We guessed the laws by looking for equations which had the greatest mathematical simplicity and beauty. Then only a few experiments were needed to test the equations and find out whether we guessed right. This happened over and over again, first with Newton's laws of motion and gravitation, then with Maxwell's equations of electromagnetism, then with Einstein's equations of special and general relativity, then with Schrödinger's and Dirac's equations of quantum mechanics. Now with string theory the game is almost over. The mathematical beauty of this theory is so compelling that it has to be right, and if it is right it explains everything from particle physics to cosmology.

Since I am reconstructing Greene's argument from memory, it is possible that I am exaggerating the claims that he was making for theoretical physics. One thing that I remember clearly is the phrase

"We are done." I still hear him saying, "We are done," in a tone of triumphant finality.

I began my reply by saying that nobody denies the amazing success of theoretical physics in the last four hundred years. Nobody denies the truth of Einstein's triumphant words: "The creative principle resides in mathematics. In a certain sense, therefore, I hold it true that pure thought can grasp reality, as the ancients dreamed." It is true that the fundamental equations of physics are simple and beautiful, and that we have good reason to expect that the equations still to be discovered will be even more simple and beautiful. But the reduction of other sciences to physics does not work. Chemistry has its own concepts, not reducible to physics. Biology and neurology have their own concepts not reducible to physics or to chemistry. The way to understand a living cell or a living brain is not to consider it as a collection of atoms. Chemistry and biology and neurology will continue to advance and to make new fundamental discoveries, no matter what happens to physics. The territory of new sciences, outside the narrow domain of theoretical physics, will continue to expand.

Theoretical science may be divided roughly into two parts, analytic and synthetic. Analytic science reduces complicated phenomena to their simpler component parts. Synthetic science builds up complicated structures from their simpler parts. Analytic science works downward to find the fundamental equations. Synthetic science works upward to find new and unexpected solutions. To understand the spectrum of an atom, you needed analytic science to give you Schrödinger's equation. To understand a protein molecule or a brain, you need synthetic science to build a structure out of atoms or neurons. Greene was saying, only analytic science is fundamental. I said, on the contrary, good science requires a balance between analytic and synthetic tools, and synthetic science becomes more and more creative as our knowledge increases.

Another reason why I believe science to be inexhaustible is Gödel's theorem. The mathematician Kurt Gödel discovered and proved the theorem in 1931. The theorem says that given any finite set of rules for doing mathematics, there are undecidable statements, mathematical statements that cannot either be proved or disproved by using these rules. Gödel gave examples of undecidable statements that cannot be proved true or false using the normal rules of logic and arithmetic. His theorem implies that pure mathematics is inexhaustible. No matter how many problems we solve, there will always be other problems that cannot be solved within the existing rules. Now I claim that because of Gödel's theorem, physics is inexhaustible too. The laws of physics are a finite set of rules, and include the rules for doing mathematics, so that Gödel's theorem applies to them. The theorem implies that even within the domain of the basic equations of physics, our knowledge will always be incomplete.

I ended by saying that I rejoiced in the fact that science is inexhaustible, and I hoped the nonscientists in the audience would rejoice too. Science has three advancing frontiers that will always remain open. There is the mathematical frontier, which will always remain open thanks to Gödel. There is the complexity frontier, which will always remain open because we are investigating objects of ever-increasing complexity, molecules, cells, animals, brains, human beings, societies. And there is the geographical frontier, which will always remain open because our unexplored universe is expanding in space and time. My hope and my belief is that there will never come a time when we shall say, "We are done."

After Greene's opening statement and my reply, the debate in Davos continued with additional remarks from us and questions from the audience. His new book and my review are a further continuation of the same debate. In the review, as in the debate, I have emphasized the points on which Greene and I disagree. There is no space here to enumerate the many points on which we agree. For both of us the most important and exciting fact is that during the last twenty years cosmology became an observational science. During the last five years, the Wilkinson Microwave Anisotropy Probe (WMAP) satellite, an orbiting radio telescope designed by my friend David Wilkinson in Princeton, has given us more detailed and precise information about the history and structure of the cosmos than all earlier telescopes combined.

Observational cosmology has now entered its golden age, with the WMAP satellite continuing to scan the sky and with a variety of even more sensitive telescopes under construction. During the next decade we shall learn far more about the cosmos than we know today, and we shall probably find new mysteries to replace those that we shall solve. Greene and I agree that so long as observers continue to explore, cosmology will continue to deepen our understanding of where we stand and how we came to be.

Postscript, 2006

After this review was published, Brian Greene wrote me a friendly letter, thanking me for the review but saying that my recollection of his remarks in the Davos debate was wrong. Since I have no wish to perpetuate errors, I deleted from this version of the review the sentences to which he objected. As a result, what is left of his remarks does not put his case forcefully. To set the record straight, here is an extract from his letter: "What I did say in Davos is that the search for the elementary ingredients making up the universe and the deepest laws governing their interactions may be a search that one day draws to a close. The deeper we look, the simpler and more unified the laws become, and there may well be a limit to this process. However, achieving this goal would only mean that we were done with one fantastically interesting but limited chapter in human exploration, the search for the basic constituents and underlying laws."

20

OPPENHEIMER AS SCIENTIST, ADMINISTRATOR, AND POET

1. Oppenheimer as Scientist

I DIVIDE THIS chapter into three parts, the first about J. Robert Oppenheimer as a scientist, the second about Robert as an administrator, the third about Robert as a poet. To make the story complete there should be a part about Robert as a statesman, but that would require another chapter as long as this one. I won't stick rigidly to these boundaries. I want to let Robert speak for himself as much as possible. The best part of the chapter will be direct quotes from Robert and others, telling us the story of his life as they saw it.

I begin in September 1938 with a story told by Robert Serber, the same Serber who appears in the movie *The Day After Trinity*. I owe this story to David Trulock, a friend of mine in Texas. The two Roberts, Serber and Oppenheimer, were at a meeting of theoretical physicists in Vancouver. The entertainment during the meeting included a boat ride among the islands offshore. The day was foggy, and navigation among the islands was done by the pilot blowing a whistle and listening for the echo. Someone asked what the consequences for physics would be if this boatload of theorists sank. Oppenheimer instantly replied, "It wouldn't do any permanent good."

One year later, on September 1, 1939, Hitler invaded Poland and started the Second World War. On that day, issue number 5 of Volume 55 of the *Physical Review* was published, containing two papers of historic importance. The first was entitled "The Mechanism of Nuclear Fission," by Niels Bohr and John Wheeler, twenty-five pages long, containing a complete and thorough theoretical explanation of the process of nuclear fission that had been discovered in Germany only nine months earlier. The second was entitled "On Continued Gravitational Contraction," by J. Robert Oppenheimer and Hartland Snyder, only four pages long, containing an equally thorough theoretical explanation of the objects that we now call black holes.

Here is the abstract of the Oppenheimer-Snyder paper, with some technical details omitted:

> When all thermonuclear sources of energy are exhausted, a sufficiently heavy star will collapse. In the present paper we study the solutions of the gravitational field equations which describe this process. The radius of the star approaches asymptotically its gravitational radius. Light from the surface of the star is progressively reddened, and can escape over a progressively narrower range of angles. The total time of collapse for an observer comoving with the stellar matter is finite, and for typical stellar masses, of the order of a day. An external observer sees the star asymptotically shrinking to its gravitational radius.

The paper is written in the same unsensational style as the abstract. Oppenheimer and Snyder did not conclude their paper by saying, "It has not escaped our notice that these collapsed objects may play a fundamental role in the dynamics and evolution of the universe," as Francis Crick and James Watson similarly said fourteen years later at the conclusion of a similar paper.

Black holes are familiar objects to modern astronomers. We know that they exist all over our own galaxy and in the central regions of other galaxies. We see them as sources of X-ray radiation emitted by gas as it falls into them and is heated to temperatures of millions of degrees by their overwhelmingly strong gravity. At the center of our own galaxy we see a black hole weighing as much as a few million suns, with massive stars orbiting around it like moths around the flame of a candle. Black holes are not rare, and they are not an accidental embellishment of our universe. They are a fundamental driving force of its evolution. They are a dominant source of energy. For every ounce of matter consumed, they yield more than ten times as much energy as the nuclear reactions of fusion and fission that cause our sun to shine and our hydrogen bombs to explode. To a modern astronomer, a universe without black holes makes no sense.

To a modern physicist, black holes are also objects of transcendent beauty. They are the only places in the universe where Einstein's theory of general relativity shows its full power and glory. Here, and nowhere else, space and time lose their individuality and merge together into a sharply curved four-dimensional structure precisely delineated by Einstein's equations. If you imagine yourself falling into a black hole, your local perception of space and time will be detached from the space and time of an observer watching you from outside. While you see yourself falling smoothly into the hole without any deceleration, the outside observer sees you coming to a halt at the horizon of the hole and remaining forever in a state of permanent free fall. Permanent free fall is a situation that can only exist by virtue of the distortion of space and time predicted by Einstein's theory.

This is the central paradox of Robert's life as a scientist. His theoretical prediction of black holes was by far his greatest scientific achievement, fundamental to the modern development of relativistic astrophysics, and yet he never showed the slightest interest in following it up. So far as I can tell, he never wanted to know whether black holes actually existed. I tried sometimes to talk with him about the possibilities for observing black holes and testing his theory. He impatiently changed the subject and talked about something else. I also met Hartland Snyder from time to time at the Brookhaven National Laboratory where he spent most of his life. He too was uninterested in black holes. He had a distinguished career as a designer of accelerators.

We now know that the Oppenheimer-Snyder calculation is correct and describes what happens to massive stars at the end of their lives. It explains why black holes are abundant, and incidentally confirms the truth of Einstein's theory of general relativity. And still, Robert was not interested. The question remains: How could he have been blind to the importance of his greatest discovery? I have no answer to this question. It remains as a paradox in the life of a genius. Perhaps if the Oppenheimer-Snyder calculation had not happened to coincide in time with the Bohr-Wheeler theory of nuclear fission and with the outbreak of World War II, Robert would have paid more attention to it.

2. *Oppenheimer as Administrator*

I do not have much firsthand experience of Robert as administrator. My chief witness here is Lansing Hammond, a friend of mine who worked for the Harkness Foundation. In 1947, when I came to the United States from England, Hammond was in charge of programs and placements for the Commonwealth Fund Fellows. In those days, Commonwealth Fund Fellows were young Brits who came to America to study at American universities with fund support. I was one of them, and Hammond made the arrangements for me to come first to Cornell University and then to the Institute for Advanced Study at Princeton. Thirty years later, in 1979, Hammond wrote me a letter about Oppenheimer. I replied to his letter, "It is sad that in the official memorials to Robert there was never anything said or written that gave such a fine impression of Robert in action. I hope there may still be a chance sometime to make your story public." Hammond died a few years later. Here is his story:

> I'd just received copies of the application papers—sixty of them—for the 1949 awards. Among them were four or five in that, to me, shadowy borderline realm between theoretical physics and mathematics. I was in Princeton for a couple of days, asking for help on all sides. Summoning all the courage I could muster, I made an appointment to see Robert Oppenheimer the next morning, leaving the relevant papers with his secretary. I was greeted graciously, asked just enough questions about my academic background to put me at ease. One early comment amazed me: "You got your doctorate at Yale in 18th century English literature—Age of Johnson. Was Tinker or Pottle your supervisor?" How did he know that?
>
> Then we got down to business. In less than ten minutes I had enough facts to support trying to persuade candidate Z that Berkeley was more likely to satisfy his particular interests than Harvard; he would fare well at the Institute; would be welcome; but Berkeley was really the best choice. I was scribbling notes as fast as I could; occasionally a proper name produced wrinkles on my forehead. Oppenheimer would flash me an understanding grin and spell out the name for me: "That may save you some time and trouble."
>
> As I was gathering up my papers, feeling I'd already taken up too much of the great man's time, he asked gently: "If you have a few minutes you can spare, I'd be interested in looking at some of your applications in other fields, to see what this year's group of young Britons are interested in pursuing over here?" I took him at his word, and was completely overwhelmed by what

ensued: "Umm—indigenous American music—Roy Harris is just the person for him, he'll take an interest in his program. Roy was at Stanford last year but he's just moved to Peabody Teachers' College in Nashville. Social psychology, he gives Michigan as first choice—Umm—he wants a general, overall experience. At Michigan he's likely to be put on a team and would learn a lot about one aspect. I'd suggest looking into Vanderbilt; smaller numbers; he'd have a better opportunity of getting what he wants." (The candidate was persuaded to try Vanderbilt for one term, with the option of transferring to Michigan if he wasn't satisfied. He spent two years at Vanderbilt, with profit and enthusiasm.) "Symbolic logic, that's Harvard, Princeton, Chicago or Berkeley; Let's see where he wants to put the emphasis. Ha! Your field, 18th-century English Lit. Yale is an obvious choice, but don't rule out Bate at Harvard, he's a youngster but a person to be reckoned with." (My field, and I'd not yet even heard of Bate, but I took pains, the next time I was in Cambridge, to meet and talk with him.) We spent at least an hour, thumbing through all of the sixty applications. Robert Oppenheimer knew what he was talking about. He pleaded ignorance about two or three esoteric programs. Every positive comment or recommendation was right on target. And so, when it finally came time to leave, I couldn't resist saying that if I could only bribe him, once a year, to repeat what he'd just done, it would save me months of sweating. He really grinned at that. "That wouldn't be fair to you, Dr. Hammond. It would take away the satisfaction and excitement of talking with lots of other people and finding out for yourself." I left, walking on air, head abuzz, most of my problems solved. Never before, never since have I talked with such a man. No suggestion of trying to impress. No need to. Robert Oppenheimer's was just genuine interest in all fields of the intellect; a fantastically up-to-date knowledge of what was going on

> in US graduate schools and research centers; an intuitive understanding of where a given person with definite interests would best fit in; and taking pleasure in being of help to someone who badly needed it.

The Robert Oppenheimer that Hammond saw that morning in 1949 was the same Robert Oppenheimer who had mastered every detail of the bomb project at Los Alamos five years earlier, and had fitted the most appropriate task to each scientist and engineer in his army of subordinates. He was equally at home in the world of literature and the world of science, in the eighteenth century and the twentieth.

The year 1942 was the turning point in Robert's life, when he suddenly changed from a left-wing academic intellectual to a practical and brilliantly successful administrator. When he accepted in 1942 the job of organizing the bomb laboratory at Los Alamos, it seemed to him natural and appropriate that he should work under the direct command of General Leslie R. Groves of the United States Army. Other leading scientists wanted to keep the laboratory under civilian control. Isidor Rabi of Columbia University was one of those most strongly opposed to working for the army. Robert wrote to Rabi in February 1943, explaining why he was willing to go with General Groves:

> I made in Washington a strong and extremely painful attempt to have our project transferred to . . . a special committee established for that purpose. I did not get to first base. . . . I do not know whether the arrangements as now outlined will work, for that will take in the first instance the good will and cooperation of quite a few good physicists, but . . . I am willing to make a faithful effort to get things going. I think if I believed with you that this project was "the culmination of three centuries of physics," I should take a different stand. To me it is primarily the development in time of war of a military weapon of some consequence.

> I do not think that the Nazis allow us the option of not carrying out that development. I know that you have good personal reasons for not wanting to join the project, and I am not asking you to do so. Like Toscanini's violinist, you do not like music.

That letter to Rabi is the only place in Robert's correspondence where he says explicitly why he pushed ahead with building the bomb and was willing to place its fate in military hands.

Late in 1944, as the Los Alamos project moved toward success, tensions developed between civilian and military participants. Captain Parsons of the US Navy, serving as associate director under Robert, complained to him in a written memorandum that some of the civilian scientists were more interested in scientific experiments than in weaponry. Robert forwarded the memorandum to General Groves, with a covering letter to show which side he himself was on: "I agree completely with all the comments of Captain Parsons' memorandum on the fallacy of regarding a controlled test as the culmination of the work of this laboratory. The laboratory is operating under a directive to produce weapons: this directive has been and will be rigorously adhered to." So vanished the possibility that there might have been a pause for reflection between the Trinity test and Hiroshima. Captain Parsons, acting in the best tradition of old-fashioned military leadership, armed the Hiroshima bomb himself and flew with it to Japan.

In later years I found a key to the character of Robert by comparing him with Lawrence of Arabia. Lawrence was in many ways like Robert, a scholar who came to greatness through war, a charismatic leader, and a gifted writer, who failed to readjust happily to peacetime existence after the war, and was accused with some justice of occasional untruthfulness. Lawrence's book *The Seven Pillars of Wisdom* is a vivid and subtly romanticized history of the Arab revolt against Turkish rule, a revolt which Lawrence orchestrated with an extraordinary mixture of diplomacy, showmanship, and military skill.

The Seven Pillars begins with a dedicatory poem, with words which perhaps tell us something about the force that drove Robert Oppenheimer to be the man he became in Los Alamos:

> *I loved you, so I drew these tides of men*
> *into my hands,*
> *And wrote my will across the sky in stars*
> *To earn you Freedom, the seven pillared*
> *worthy house,*
> *That your eyes might be shining for me*
> *When we came,*

and with words which tell of the bitterness which came to him afterward:

> *Men prayed that I set our work, the inviolate*
> *house,*
> *As a memory of you.*
> *But for fit monument I shattered it, unfinished:*
> *and now*
> *The little things creep out to patch themselves*
> *hovels*
> *In the marred shadow*
> *Of your gift.*

3. Oppenheimer as Poet

Robert Oppenheimer was also something of a poet. The best place to find the poetic Robert is in the book *Robert Oppenheimer: Letters and Recollections*,[1] a collection of personal letters, with recollections

1. Edited by Alice K. Smith and Charles Weiner (Harvard University Press, 1980).

contributed by his friends and recorded by the editors Alice Smith and Charles Weiner. I quote three snippets to give you a taste of the young Robert. The first was written when he was a nineteen-year-old sophomore at Harvard, to Miss Limpet (actually his boyhood friend Paul Horgan) from Celia (himself) describing the antics of Celia's son Henley (also himself). The letter ends with a parody by Henley of Eliot's recently published *The Waste Land*:

> *What does it mean?*
> *Decaying hag*
> *Shrewish in the wilted sheen*
> *Of a stoop; raucous stag*
> *Boasting loot in flesh and waistcoat,*
> *Pretty penny, Henley, Ascot,*
> *Left 'em not a rag.*
> *No, that is not what it means.*

Four years later Robert was back at Harvard as a postdoc, having finished his Ph.D. with Max Born at Göttingen in record time. At the age of twenty-three he published a poem of his own with the title "Crossing." It describes the landscape of New Mexico which he had come to love:

> *It was evening when we came to the river*
> *With a low moon over the desert*
> *That we had lost in the mountains, forgotten,*
> *What with the cold and the sweating*
> *And the ranges barring the sky.*
> *And when we found it again,*
> *In the dry hills down by the river,*
> *Half withered, we had*
> *The hot winds against us.*

There were two palms by the landing;
The yuccas were flowering; there was
A light on the far shore, and tamarisks.
We waited a long time, in silence.
Then we heard the oars creaking
And afterwards, I remember,
The boatman called to us.
We did not look back at the mountains.

My third snippet comes from a letter of Robert to his brother Frank, written when Robert was twenty-eight, teaching physics and building a first-rate school of research in California. Frank was eight years younger and was then an undergraduate at Johns Hopkins. Here is some of Robert's fatherly advice:

> The fact that discipline is good for the soul is more fundamental than any of the grounds given for its goodness.... But because I believe that the reward of discipline is greater than its immediate objective, I would not have you think that discipline without objective is possible: in its nature discipline involves the subjection of the soul to some perhaps minor end; and that end must be real, if the discipline is not to be factitious. Therefore I think that all things which evoke discipline: study, and our duties to men and to the commonwealth, war, and personal hardship, and even the need for subsistence, ought to be greeted by us with profound gratitude; for only through them can we attain to the least detachment; and only so can we know peace.

It comes as a shock to see that little word "war," among the things that we should be grateful for. This may help to explain how easily Robert slipped into the role of the good soldier ten years later.

These letters give some insight into Robert's character, showing us the flaw which made his life ultimately tragic. His flaw was restlessness, an inborn inability to be idle. Intervals of idleness are probably essential to creative work on the highest level. Shakespeare, we are told, was habitually idle between plays. Robert was hardly ever idle. His restlessness appears already in the early Harvard letters, outpourings of words written by a young man unable to stop when he has nothing more to say. Restlessness was at the root of the craving for discipline that is revealed in his letters to his brother. Restlessness drove him to his supreme achievement, the fulfillment of the mission of Los Alamos, without pause for rest or reflection. Without his restlessness, the pace at Los Alamos would have been slower. There would then have been a chance for the Second World War to have ended quietly in a Japanese surrender with Hiroshima and Nagasaki spared.

Robert was well aware of his own weakness. In later life he never spoke of himself directly, but he occasionally expressed his inner thoughts obliquely by quoting poetry. Especially from George Herbert, his favorite poet. In my Oppenheimer file there is a letter from Ursula Niebuhr, who knew Robert better than I did. She was the wife of the famous theologian Reinhold Niebuhr, who was invited to the institute by Robert and lived here as a member. Here is Ursula writing:

> The last comment is about George Herbert. It was another lunchtime. This one was at the Oppenheimers' house, on a beautiful spring day, and Kitty had masses of daffodils about the house. The Kennans and we were invited. Robert was at his most charming and hospitable best. After lunch, over coffee in that old part of their living room on the lower level, with Robert's favorite books in the black-painted bookcases at the back, and the sunlight on the daffodils, and the smell of the wood fire, somehow Robert discovered that George Kennan did not know George Herbert. He turned to me and said, "But you,

of course, do." My father had been named for George Herbert, as there had been some distant connection over two hundred years ago, at least according to my pious grandmother. Robert went to his bookcase and drew out a rather nice old edition of Herbert and read in that sympathetic voice of his "The Pulley":

> *When God at first made man,*
> *Having a glasse of blessings standing by,*
> *"Let us," said He, "poure on him all we can;*
> *Let the world's riches, which dispersèd lie*
> *Contract into a span."*

and ending as no doubt you recall with the lines:

> *Yet let him keep the rest,*
> *But keep them with repining restlessnesse;*
> *Let him be rich and wearie, that at least,*
> *If goodnesse leade him not, yet wearinesse*
> *May tosse him to My breast.*

Robert then said, "Well, we've got to see that George Kennan reads George Herbert."

When Robert died in 1967, his wife, Kitty, called me in to discuss the arrangements for the memorial ceremony. Besides the music and the talks by Robert's friends describing his life and work, she wanted also to have a poem read, since poetry had always been an important part of Robert's life. She knew which poem she wanted to have read, "The Collar" by George Herbert, a poem that had been one of Robert's favorites. She found it particularly appropriate to describe how Robert had appeared to himself. Then she changed her mind. "No," she said, "that is too personal for such a public occasion." She had good reason for being afraid to bare Robert's soul in public. She knew from bitter experience how newspapers are apt to handle such disclosures. She could already imagine the horrible distortions of Robert's true feelings, appearing under the headline "Noted Scientist, Father of Atom Bomb, Turns to Religion in Last Illness." No poem was read at the ceremony.

21

SEEING THE UNSEEN

EVERY ATOM IS almost entirely made of empty space, with a tiny object called the nucleus and even tinier objects called electrons flying around inside it. Ernest Rutherford, a young New Zealander working in Manchester, England, discovered this fact about atoms in 1909. He shot fast particles at a thin film of gold and observed the way the particles bounced back. The pattern of the recoiling particles showed directly the internal structure of the atoms in the film. The discovery of the tiny nucleus came as a big surprise to Rutherford as well as to everybody else. The phrase "the fly in the cathedral" described what Rutherford discovered. The fly is the nucleus; the cathedral is the atom. Rutherford's experiment showed that almost all the mass and almost all the energy of the atom was in the nucleus, although the nucleus occupied less than a trillionth part of the volume.

Rutherford's discovery was the beginning of the science that came to be called nuclear physics. After discovering that nuclei of atoms exist, he continued to study their properties by bombarding them with fast particles and observing the results. The projectiles that he used to explore the nucleus were particles produced in the disintegration of radium. Radium is a naturally occurring radioactive metal that was discovered by Marie Curie in 1898. The particles are helium nuclei that are emitted at high speed when radium atoms decay. These particles made good probes for exploring the properties of nuclei because they were all alike and came with known energies. For twenty years, first in Manchester and later in Cambridge, Rutherford and his students and colleagues used natural particles with great success to learn how nuclei behave. They found that it was possible on rare occasions to change one kind of nucleus into another by adding or subtracting a particle. The twenty years between 1909 and 1929 were the era of tabletop nuclear physics. Experiments were small enough to fit onto the tops of tables. Small and simple experiments were sufficient to establish the basic laws of nuclear physics.

Toward the end of the 1920s, nuclear physics got stuck. Major mysteries remained to be solved. Nobody knew what nuclei were made of or how their component parts were put together. But it was hard to think of exciting new experiments that could be done with the existing tools. The next round of experiments were minor variations of experiments that had already been done. It seemed unlikely that the mysteries of nuclear structure could be solved by such experiments. Rutherford announced in a public lecture in London in 1927 that new tools were needed if nuclear physics were to move ahead. Without new tools, research in nuclear physics would stagnate and bright young people would no longer be attracted to it. The most promising new tool would be a particle accelerator, an electrical machine that could produce a beam of artificially accelerated particles to replace the natural particles produced by radium. Artificially accelerated particles would be better than natural particles in three ways. They could be produced in greater quantities, they could have higher energies, and they would allow experiments to be designed more flexibly. The switch from natural sources of particles to accelerators would start a new era in the history of science, the era of accelerator physics.

The Fly in the Cathedral[1] tells how the era of accelerator physics

1. Brian Cathcart, *The Fly in the Cathedral: How a Group of Cambridge Scientists Won the International Race to Split the Atom* (Farrar, Straus and Giroux, 2004).

started. It is a dramatic story and Brian Cathcart tells it well. He is a journalist and not a scientist, but he understands enough of the science to get the details right. He has made a thorough study of the primary literature, he has read the papers and letters written by the participants, and he interviewed those of them who were still alive. The story begins with Rutherford's decision in 1927 to explore the possibility of accelerating particles, and ends with the building of the first accelerator and its triumphant success as an atom-splitter in 1932. The era of accelerator physics began in 1932 and is not yet ended. The enormously powerful accelerators that are now exploring the fundamental forces of nature in Illinois and California and Switzerland and Japan are the direct descendants of machines that were built in 1932.

The story of the first accelerator was not only an important chapter in the history of science. It was also an international sporting event, driven by national pride as well as by scientific curiosity. Rutherford had competitors in many countries. The most formidable competitors were in the United States: Merle Tuve at the Carnegie Institution in Washington, Robert Van de Graaff at the Massachusetts Institute of Technology, and Ernest Lawrence at the University of California in Berkeley. Rutherford knew and respected his competitors but was determined to beat them. As a scientist he was a member of an international community, but as an old-fashioned New Zealander he was fiercely loyal to Britain and the Empire. He understood that science is an international enterprise that flourishes best when nations compete for the prizes.

The two men who actually built the first accelerator were John Cockcroft and Ernest Walton, graduate students working in the Cavendish Laboratory in Cambridge under the supervision of Rutherford. Cockcroft came to the Cavendish from Yorkshire in 1924, Walton from Ireland in 1927. When Walton arrived from Dublin, he proposed to Rutherford for his student project that he should start building an accelerator, not knowing that Rutherford had already announced his intention to do so. Rutherford was happy to give him the green light. Cockcroft had worked at Metropolitan-Vickers, the leading electric engineering company in Britain, before coming to Cambridge, so that he had some experience in working with heavy equipment and high voltages. Rutherford arranged for Walton to work full-time on the accelerator project, with Cockcroft working on it part-time to help with the engineering. For five years they struggled to create a technology of big machines in a laboratory of tabletop experiments, just as the Wright brothers had struggled to create a technology of flying machines in a bicycle shop.

The obstacles that Cockcroft and Walton had to overcome were cultural as well as technical. If they were to build big machines, they would obviously need more space, but the idea of building a new building to house a new machine was not thinkable in the Cavendish culture. Rutherford was notoriously stingy and kept all expenditures to a minimum. The Cavendish was a historic building and could not be remodeled. So everything that Cockcroft and Walton built had to fit into existing rooms and pass through existing doors. This had the consequence that they could not use any commercial power equipment that could not be squeezed through the historic Gothic gateway of the Cavendish. They had to spend long months designing and testing their own equipment.

The culture of the Cavendish was strongly paternalistic. Rutherford took fatherly care of his students and imposed strict limits on their hours of work. Every evening at six o'clock the laboratory was closed and all work had to stop. Four times every year, the laboratory was closed for two weeks of vacation. Rutherford believed that scientists were more creative if they spent evenings relaxing with their families and enjoyed frequent holidays. He was probably right. Working under his rules, an astonishingly high proportion of his students, including Cockcroft and Walton, won Nobel Prizes.

They were maintaining the culture of nineteenth-century gentlemen-scientists, who were supposed to pursue scholarly leisure-time activities in addition to their science. But this culture of short hours and plentiful leisure was hard to combine with heavy machine–building. Years went by while Cockcroft and Walton put together improvised vacuum systems, laboriously sealed the leaks, tried out various devices for handling high voltages, and found them to be inadequate.

It took Cockcroft and Walton five years to build a machine that worked. Finally, in April 1932, they had a machine that produced a steady stream of hydrogen nuclei with an energy of about half a million volts. They were carefully measuring the quality of the stream, in no hurry to begin doing nuclear experiments. On the morning of April 13, Rutherford came into their room, saw what they were doing, and lost his temper. He told them to stop wasting their time and do some science. The next day, Cockcroft was occupied with other business. Walton was alone in the laboratory, ready to do the first experiment, using his hydrogen nuclei to bombard a target made of the light metal lithium. The result was spectacular. The lithium nuclei were split in two and fell apart into pairs of helium nuclei. The helium nuclei came out with thirty times as much energy as the hydrogen nuclei going in. Walton ran to Rutherford's office to tell him the news, and Rutherford happily spent the rest of the day serving as Walton's assistant, checking the result and tidying up the details. That day, the era of tabletop nuclear physics ended and the era of big machines and big projects began.

The American competitors were running close behind Rutherford and were beaten only by a few weeks. Van de Graaff had invented an electrostatic accelerator that was in many ways superior to the Cockcroft-Walton machine, and Lawrence had invented the cyclotron, which was in many ways better still. They were not forbidden to work after six o'clock in the evening, but they, too, had to struggle with cultural obstacles similar to those that hampered Cockcroft and

Walton. American academic administrators, hard hit by the economic depression of the 1930s, were almost as stingy as Rutherford. Tuve and Van de Graaff had built a splendid machine at the Carnegie Institution in Washington, but it was standing on the grass out-of-doors, where dust and insects interfered with its operation. They were forced to dismantle it because the laboratory had no room large enough to house it.

Even Lawrence, who became in later years the main driving force of big-machine physics, had similar problems when he started to build cyclotrons. In 1931 he had acquired for his newest cyclotron a huge magnet that weighed eighty-five tons, but he had no building big enough for it and so the cyclotron remained unfinished. Van de Graaff and Lawrence were the hares, Rutherford was the tortoise, and the tortoise won the race.

In the few years that remained to him after 1932, Rutherford enjoyed the rebirth of nuclear physics brought about by the new machines. His students in the Cavendish continued to explore the universe of nuclei, in friendly competition with explorers in America and Europe. He died a year before the discovery of the fission of the uranium nucleus in Berlin in 1938, the discovery which turned nuclear physics into a big industry and a weapon of war. Walton returned to Dublin in 1934 and spent the rest of his life there peacefully as a professor of physics. Cockcroft stayed at the Cavendish until 1939, then moved into war research, and then became director of the British Atomic Energy Research Establishment at Harwell when it was established after the war. Harwell was mainly concerned with the development of reactors, for scientific research and for the nuclear power industry. Cockcroft showed me around Harwell when I visited there in the 1950s. The most conspicuous feature of Harwell at that time was a huge array of high-voltage cables connecting the laboratory to the national electricity grid. "The main reason the public supports us," Cockcroft said with a smile, "is that they think the electricity is flowing out of the lab. Actually, of course, it is flowing in."

* * *

Cathcart's book ends with a discussion of the question why the tortoise beat the hares in the race to disintegrate nuclei with accelerators. He concludes that the main reason why Rutherford won was that he was not a machine builder. Van de Graaff and Lawrence were brilliant inventors, driven by passionate love for their machines. They were not so much concerned with what their machines could do when they were built. For Rutherford the machine was only a tool. He was not interested in the details of its design and trusted Cockcroft and Walton to get the details right. For him, what mattered was the science.

He had spent his life exploring the nuclei of atoms, and his driving passion was to dig deeper into the nucleus. That was why he made sure that the decisive experiment was carefully prepared and the lithium nuclei were ready and waiting to be disintegrated as soon as his machine was finished. The Americans had better machines, but Rutherford had a more single-minded concentration on the scientific goal. Two months after he won the race, Rutherford explained to a reporter from the *Daily Herald* why he had wanted so badly to disintegrate nuclei. "We are rather like children," said Rutherford, "who must take a watch to pieces to see how it works."

Alan Lightman's book *A Sense of the Mysterious*[2] tells a very different story. His book has no index, so I cannot be sure how many times Rutherford is mentioned. I believe he is mentioned only once, on page 133, in a list of names of famous people who were dead in the year 2002 when that page was written. I find it remarkable that we have two books designed to give nonexpert readers a feeling for the way research into the nature of the physical universe is done, and yet the central character of one book is barely mentioned in the other. How could two accounts of the same science be so different, in so many ways, in style and in substance?

2. Pantheon, 2005.

Brian Cathcart is an Irish journalist with an amateur interest in science. Alan Lightman is an American, trained as a theoretical physicist, who made a midlife change of career from scientific research to writing. Cathcart's book is a straightforward historical narrative. Lightman's book is a collection of essays, lectures, and book reviews, most of them describing individual scientists and their ideas. Cathcart is primarily interested in experiments, Lightman in theories. Cathcart sees progress in science mainly driven by new tools; Lightman sees progress driven by new concepts. Cathcart's story is a simple drama with three heroes and no villains, a triumph of human pertinacity over technical and cultural obstacles. Lightman's chapters are meditations on the human condition, illustrated by sketches of characters who are partly heroes and partly villains.

The main characters in Lightman's stories are the theoretical physicists Albert Einstein, Edward Teller, and Richard Feynman, and the observational astronomer Vera Rubin. Not only is Rutherford absent, but almost all experimental scientists are absent too. The only experimenter who makes an appearance on Lightman's stage is Joseph Weber, a brilliant and tragic figure whose experiments turned out to be wrong. The mainstream experimenters who explored the universe of particles and fields, continuing to play Rutherford's game, taking the watch to pieces to see how it works, do not appear at all. Lightman's title, *A Sense of the Mysterious*, and his subtitle, *Science and the Human Spirit*, do not explain his neglect of experimenters. After all, Rutherford had as deep a sense of the mysteries of nature as Einstein. And the human spirit expresses itself as eloquently in the work of human hands as in the work of human minds. Rutherford was supreme as an experimenter and Einstein was supreme as a theorist, but each of them held the other in deep respect. Both of them understood that the human spirit is at its best when hands and minds are working together.

One theorist played an essential role in Rutherford's thinking. George Gamow was a brilliant young Russian who came to Germany in 1928 and at the age of twenty-four started a revolution in nuclear physics. He was the first to understand how the quantum theory, which had been invented only three years earlier, could be applied to the nucleus. He used quantum theory to calculate how fast radioactive nuclei such as radium or uranium should decay, and found that the theory agreed pretty well with the observed rates of disintegration. He then made another decisive step, using quantum theory to calculate how easily a charged particle could come into a nucleus from outside. He understood that the same quantum rules apply in both directions. Easy out, easy in. If particles can escape from radioactive nuclei by quantum rules, then they can also penetrate into nuclei by quantum rules when fired at them from outside.

When Rutherford heard of Gamow's idea, he saw at once that this dramatically improved the prospects for doing important science with the accelerator that he was planning to build. Rutherford did not pretend to understand quantum mechanics, but he understood that the Gamow formula would give his accelerator a crucial advantage. Even particles accelerated to much lower energies than the particles emitted naturally by radium would be able to penetrate into nuclei. Rutherford invited Gamow to Cambridge in January 1929. The fifty-eight-year-old experimenter and the twenty-four-year-old theorist became firm friends, and Gamow's insight gave Rutherford the impetus to go full steam ahead with the building of his accelerator.

The same mutual admiration of experimenter and theorist was shown three years later when Einstein happened to be visiting Cambridge a few days after the triumph of Cockcroft and Walton. Einstein insisted on seeing the accelerator that had split the atom. Walton spent a morning showing him the apparatus and explaining the details of its operation. Einstein wrote a letter afterward, expressing "astonishment and admiration" for what he had seen. "He seems a very nice sort of man," wrote the imperturbable Walton to his fiancée in Ireland.

How is it that this mutual admiration and easy mixing of theory with experiment, which seemed natural and necessary in the 1930s, is absent from Lightman's view of physics? Somehow it happened that the successors of Rutherford and Einstein drifted apart in the second half of the twentieth century. This was not the physicists' fault. It resulted from the enormous growth of accelerators and the enormous proliferation of theories. Accelerators and the accompanying apparatus for detecting particles became so huge and complicated that each experiment was like a military operation. Hundreds of people with highly specialized skills were required to carry out a program planned many years in advance. Theorists became similarly specialized, some of them expert in accelerator design, some in particle interactions, some in general relativity, and some in string theory. It became difficult for theorists in different specialties to communicate with one another, let alone with experimenters. At the end of the century, accelerator physics was slowing down. Each experiment required about a decade to design and prepare. Lightman, an imaginative theorist who liked to avoid narrow specialization, found such experiments unattractive. It was natural for him, following his sense of the beautiful, to move away from experimental physics and toward astronomy.

Astronomers have so far escaped the extreme specialization that has overtaken physicists. Telescopes are big, but they are not as complicated as accelerators. Observations with a big telescope can be carried out in hours rather than years. Astronomers can be skilled observers and also expert in the theory of what they are observing. That is why the astronomer Vera Rubin has a place of honor in Lightman's book. She started her professional career as a student of George Gamow after Gamow moved to America. She spent the rest of her professional life observing galaxies and studying their dynamics. She found that the visible matter in galaxies is not heavy enough to explain the speed of their internal motions. She deduced from her observations that galaxies are pervaded by dark matter, invisible to our telescopes. Nobody knows what dark matter is. It is another deep mystery remaining to be explored. We know only that it is there, and that it weighs more than all the stuff that we can see.

Besides discovering and exploring dark matter, Rubin raised four children and crusaded publicly for the advancement of women in science. I was recently chairman of a committee that organized a scientific conference with a list of distinguished scientists as members. I received a blistering letter from Rubin, asking why we had no women on our list. She supplied me with another list of women who should have been invited. I wrote back to apologize and to thank her for her list, which I shall certainly use in the unlikely event that I ever become chairman of another such committee.

Lightman's chapter on Edward Teller is a review of Teller's memoirs. Lightman considers Teller to be on the whole an evil character, in sharp contrast to his sympathetic portrayals of Einstein and Feynman. The title of the Teller chapter is "Megaton Man," emphasizing the obsession with hydrogen bombs which made Teller famous. Lightman admits that there were two Tellers. He writes, "There is a warm, vulnerable, honestly conflicted, idealistic Teller, and there is a maniacal, dangerous, and devious Teller." But his portrait of Teller shows us mostly the dark side. I knew Teller well and worked with him joyfully for three months on the design of a safe nuclear reactor. The Teller that I knew was the warm, idealistic Teller. We disagreed fiercely about almost everything and remained friends. He was the best scientific collaborator I ever had. I consider Lightman's portrayal of him to be unjust. My own review of Teller's memoirs explains why.[3]

---

3. See Chapter 15. Edward Teller describes in his memoirs the only time he met Rutherford. Rutherford gave in 1934 a lecture denouncing as a lunatic anyone who imagined that nuclear energy might ever be put to practical use.The lecture was given in London, soon after Teller arrived in England as a refugee. Teller was in the audience and was not favorably impressed. He afterward learned that the lunatic who had aroused Rutherford's anger was his friend Leo Szilard. Szilard had tried unsuccessfully to persuade Rutherford that a neutron chain reaction was a practical and dangerous possibility. It is interesting to speculate how different the history of the last century might have been if Rutherford had taken Szilard's warning seriously.

Putting together the portrait of Rutherford in Cathcart's book with my own recollections of Teller, I find striking similarities. Rutherford and Teller were both immigrants who became fiercely patriotic in defense of their adopted countries. Both often behaved like overgrown children, losing their tempers over trivialities and then regaining their equilibrium with a friendly smile. Both were father figures to their students, taking care of students' personal problems as well as their professional education. Both were more interested in the strategy of science than in the tactics. Rutherford made the decision to explore nuclei with an accelerator, and then left the details of the accelerator to Cockcroft and Walton. Teller made the decision to build a hydrogen bomb or a safe reactor and then left the details to others. Both had a lifelong dedication to science, but spent more time helping younger people than doing research themselves. Teller published his version of the hydrogen bomb story under the title *The Work of Many People*. The names of Cockcroft and Walton appear on the letter to *Nature* announcing their discovery but Rutherford's does not. My name appears on the patent for the safe reactor but Teller's does not.

The most concise and original chapter in Lightman's book is "Metaphor in Science," an essay originally published in 1988 in *The American Scholar*. Illustrating his thesis with quotations from great physicists from Isaac Newton to Niels Bohr, Lightman traces the powerful influence of metaphors on their thinking. As science has become more abstract and remote from everyday experience, the role of metaphor in our descriptions of the world has become more central. The language that nature speaks, as Galileo long ago pointed out, is mathematics. The language that ordinary human beings speak, especially those of us who are not fluent in mathematics, is metaphor. Lightman ends his discussion with another metaphor: "We are blind men, imagining what we don't see." That is a good description of theoretical physics.

22

THE TRAGIC TALE OF A GENIUS

NORBERT WIENER WAS famous at the beginning of his life and at the end. For thirty years in the middle during which he did his best work, he was comparatively unknown. He was famous at the beginning as a child prodigy. His father, Leo Wiener, the first Jew to be appointed a professor at Harvard, was a specialist in Slavic languages. Leo was also an extreme example of a pushy parent. He drove Norbert unmercifully, schooling him at home in Greek, Latin, mathematics, physics, and chemistry. Fifty years later Norbert described, in his autobiography, *Ex-prodigy: My Childhood and Youth*,[1] how the prodigy was nurtured:

> He would begin the discussion in an easy, conversational tone. This lasted exactly until I made the first mathematical mistake. Then the gentle and loving father was replaced by the avenger of the blood.... Father was raging, I was weeping, and my mother did her best to defend me, although hers was a losing battle.

At age eleven, Leo enrolled Norbert as a student at Tufts University, where he graduated with a degree in mathematics at age fourteen.

1. Simon and Schuster, 1953.

Norbert then moved to Harvard as a graduate student and emerged with a Ph.D. in mathematical logic at age eighteen. While he was growing up and trying to escape from his notoriety as a prodigy at Tufts and Harvard, Leo was making matters worse by trumpeting Norbert's accomplishments in newspapers and popular magazines. Leo was emphatic in claiming that his son was not unusually gifted, that any advantage that Norbert had gained over other children was due to his better training. "When this was written down in ineffaceable printer's ink," said Norbert in *Ex-prodigy*, "it declared to the public that my failures were my own but my successes were my father's."

Miraculously, after ten years of Leo's training and seven years of tortured adolescence, Norbert settled down to adult life as an instructor at the Massachusetts Institute of Technology and became a productive mathematician. He climbed the academic ladder at MIT until he was a full professor, and stayed there for the rest of his life. For thirty years, roughly from age twenty to age fifty, he faded from public view. He remained famous in the MIT community for his personal eccentricities. He liked to think aloud and needed listeners to hear what he was thinking. He made a habit of wandering around the campus and talking at great length to any colleague or student that he encountered. Most of the time, the listeners had only a vague idea of what he was talking about. Colleagues and students who valued their time learned to hide when they saw him coming. At the same time, they respected him for his achievements and for his encyclopedic knowledge of many subjects.

Wiener was unusual among mathematicians in being equally at home in pure and applied mathematics. He made his reputation as a pure mathematician by inventing concepts such as "Wiener measure" that have passed into the mainstream of mathematics. Wiener measure gave mathematicians for the first time a rigorous way to talk about the collective behavior of wiggly curves or flexible surfaces. While continuing to publish papers in the abstract realms of mathematical logic and analysis, he loved to talk with the engineers and neurophysiologists who were his neighbors at MIT and Harvard. He became deeply immersed in their cultures, and enjoyed translating problems from the languages of engineering and neurophysiology into the language of mathematics.

Unlike most pure mathematicians, he did not consider it beneath his dignity to apply his skills to the messy practical problems of the real world. He became a successful applied mathematician, helping to design machines and communication systems for use in war and peace. He understood, more clearly than anyone else, that the messiness of the real world was precisely the point at which his mathematics should be aimed. As an applied mathematician, he worked out a general theory of control systems and feedback mechanisms, a theory which he called "cybernetics." Cybernetics was a theory of messiness, a theory that allowed people to find an optimum way to deal with a world full of poorly known agents and unpredictable events. The word "cybernetics" comes from the Greek word for steersman, the man who steers a frail ship through stormy seas between treacherous rocks.

During World War II, Wiener worked with his engineer friend Julian Bigelow designing an optimum control system for antiaircraft guns. The design of the control system was an elementary exercise in cybernetics. Like his colleagues at MIT, Wiener was happy to be engaged in work that could help to win the war. To shoot down an airplane, it was necessary to predict the future position of the airplane at the time of arrival of the shell, knowing only the past track of the airplane up to the moment when the prediction was made. During the interval between prediction and arrival, the pilot of the airplane would be taking evasive action, changing his course in a way that could only be estimated statistically. To maximize the chance of destroying the airplane, the control system must take into account the multitude of wiggly paths that the airplane might follow. The concept of Wiener measure was the tool that allowed him to translate the problem of finding an optimum prediction into precise mathematical language. He worked hard with Bigelow to translate the mathematical solution of the problem back into electrical and mechanical hardware. Unfortunately, the United States Army could not wait for the Wiener-Bigelow hardware to be manufactured and tested. The army needed an antiaircraft control system that could be mass-produced and deployed on the battlefield as soon as possible. The army chose a less sophisticated control system that would be available sooner, designed by a rival group of engineers at the Bell Laboratories.

The Bell system became operational and the Wiener-Bigelow system never saw combat. In the end, the choice of the Bell system probably had little effect on the course of the war. The big breakthrough in antiaircraft technology was the invention of the proximity fuse, a radar-controlled fuse that enabled a shell to explode and destroy an airplane nearby without directly hitting it. Without proximity fuses, neither the Bell system nor the Wiener-Bigelow system was accurate enough to shoot down airplanes reliably. After proximity fuses became available in 1944, the Bell system was good enough.

When the war ended with the nuclear attacks on Hiroshima and Nagasaki in 1945, Wiener was outraged. In his eyes, the government had committed a crime against humanity, and the scientists who had created the bomb were to blame for allowing the government to exploit their skills for evil purposes. The nuclear attacks confirmed a belief that had been growing in his mind for many years, that the technology of communication and control which he had helped to create was fundamentally dangerous. He saw the nuclear attacks as a glaring example of the disasters that could result from science and technology when scientists were working in secrecy for military and industrial authorities. He feared that the nascent technology of computers and automatic machinery could lead to even greater disasters if it remained in the hands of secret military and industrial organizations. He decided from that moment on to have nothing to do either with government or with industry. He decided to devote a major part of his time to educating the public, to helping it deal wisely with new technologies.

In January 1947, Wiener published in *The Atlantic Monthly* an article with the title "A Scientist Rebels," an eloquent statement of his refusal to cooperate with the government. "I do not expect to publish any future work," he wrote, "which may do damage in the hands of irresponsible militarists." This article immediately made him as famous at the age of fifty-two as he had been as a child prodigy. For the rest of his life, he continued to be well known as a political activist, writing articles and books that were widely read, traveling to many countries to meet with political leaders and concerned citizens. As he explained in his second autobiography, *I Am a Mathematician: The Later Life of a Prodigy*,[2] "I thus decided that I would have to turn from a position of the greatest secrecy to a position of the greatest publicity, and bring to the attention of all the possibilities and dangers of the new developments."

For the last decade of his life, Wiener was a prophet who spoke and wrote eloquently about the displacement of human beings by automatic machinery. He saw this displacement as a likely consequence of his own inventions. But he spoke and wrote with equal eloquence of the good that automatic machinery could do, if it were used intelligently to make poor societies rich, to enable poor countries to jump from an agricultural economy to an industrial economy without enduring the horrors of nineteenth-century industrialization. He published two books that became best sellers, *Cybernetics; or, Control and Communication in the Animal and the Machine*, in 1948,[3] and *The Human Use of Human Beings: Cybernetics and Society*, in 1950.[4]

2. Doubleday, 1956.

3. John Wiley.

4. Houghton Mifflin.

Before modern electronic computers existed, these books predicted with some degree of accuracy the economic and political effects of computer technology on human societies. "We were here," he wrote,

> in the presence of another social potentiality of unheard-of importance for good and for evil.... It gives the human race a new and most effective collection of mechanical slaves to perform its labor.... However, any labor that accepts the conditions of competition with slave labor accepts the conditions of slave labor, and is essentially slave labor.... The answer, of course, is to have a society based on human values other than buying or selling.

He concluded his sermon with a sentence borrowed from the medieval poet Bernard of Cluny, "The hour is very late, and the choice of good and evil knocks at our door."

Wiener shared the fate of other major prophets, being honored abroad more than at home. He was honored most spectacularly in India and Russia. He traveled several times to India and was welcomed personally by Nehru and other Indian leaders. He went on lecture tours, and gave advice about industrial policy to the Indian government. He advocated the founding of technical institutes and the encouragement of homegrown technical industries. His advice is bearing fruit fifty years later, as India emerges as a major center of information technology and American business is outsourced to Indian firms. He also traveled to Russia, where he received equally strong official adulation but felt less at home. He told the Russians that science must be free from the restraints of political ideology. He found the ideology of Marxism as destructive of human values as the ideology of free-market capitalism. The Soviet government ignored his plea for scientific freedom but enthusiastically supported cybernetics.

The cult of cybernetics in Russia was more philosophical than practical, but it may have had some lasting effects. Perhaps it contributed to the recent emergence of a computer-literate society and a homegrown software industry. In 1964, at the age of sixty-nine, Wiener was invited to give lectures about cybernetics in Sweden, where his ideas also had a wide following. The day after his arrival, he died suddenly of a pulmonary embolism on the steps of the Royal Institute of Technology in Stockholm.

Dark Hero of the Information Age[5] is the third biography of Norbert Wiener, unless there are others of which I am ignorant. First came a joint biography of Wiener and the mathematician John von Neumann, *John von Neumann and Norbert Wiener: From Mathematics to the Technologies of Life and Death*, by Steve Heims in 1980.[6] Then came *Norbert Wiener, 1894–1964*, by Pesi Masani in 1990.[7] The main justification for a new biography is that the three biographies emphasize different aspects of Wiener's life and character. The Heims biography emphasizes politics. It is mainly concerned with Wiener's activities as a social critic in the last third of his life. It presents the parallel lives of von Neumann and Wiener as a simple struggle between black and white, with von Neumann as the evil genius of science in the service of war, and Wiener as the good genius of science in the service of peace.

The authors of the new biography cite Heims frequently, but do not accept his judgments uncritically. They present the relationship between von Neumann and Wiener as it appears in the historical documents, a friendship based on common interests and deep mutual

---

5. Flo Conway and Jim Siegelman, *Dark Hero of the Information Age: In Search of Norbert Wiener, the Father of Cybernetics* (Basic Books, 2005).

6. MIT Press.

7. Birkhäuser.

respect. The paths of von Neumann and Wiener diverged after World War II when von Neumann was willing to accept money from the United States government to support his research and Wiener was not. After that, they saw little of each other, but the mutual respect endured. When von Neumann started building a digital computer in Princeton in 1946, Wiener recommended his collaborator Julian Bigelow to be in charge of the hardware, and Bigelow became von Neumann's chief engineer with Wiener's blessing.

Pesi Masani's biography is from a scholarly point of view the best of the three. Masani was a professional mathematician, born in India and settled in the United States. He collaborated with Wiener and published several substantial papers with him in the 1950s. After Wiener died, Masani edited his collected papers for publication. Masani was intimately familiar with every detail of Wiener's work. The Masani biography is the only one that portrays him as a working mathematician. Any biography that skips the mathematics can give only a vague impression of Wiener's way of thinking. Masani states his purpose at the beginning of his book: "This book attempts to trace the interaction between mathematical genius and history that has led to the conception of a stochastic cosmos."

Masani explains Wiener's mathematical ideas with admirable clarity, and he has found and reproduced many historical documents that the other biographers have missed. One particularly illuminating document that Masani reproduces in full is a long and friendly letter from von Neumann to Wiener, written in November 1946, discussing the mysteries of the human brain and the various ways in which the mysteries might be explored. "I am most anxious to have your reaction to these suggestions," von Neumann writes. "I feel an intense need that we discuss the subject extensively with each other." Von Neumann's letter shows how far he had come in foreshadowing the era of molecular biology that he never lived to see. Von Neumann and Wiener shared a passionate interest in biology. Both of them saw a deeper understanding of biology as the ultimate goal of their explorations of the science of computing and information.

After Heims has described Wiener's politics and Masani has described his mathematics, what is there left for a third biography to do? This third biography gives us a new and intimate portrait of Wiener as a person, and describes his stormy relationships with his friends and family. Flo Conway and Jim Siegelman have done a thorough job of historical research, interviewing most of the surviving witnesses, and documenting the narrative with detailed references to published and unpublished papers, letters, and interviews. The title, *Dark Hero of the Information Age*, indicates their main preoccupation. Their aim is to explore the roots of Wiener's lifelong malaise and often weird behavior. Their intimate portrait became possible because they enjoyed the cooperation of Wiener's daughters, Barbara and Peggy, who gave them free access to Wiener's private papers and family records. Peggy wrote in a letter, "Serious unanswered questions remain concerning Dad's life and relations with his colleagues. It is very important to tell the whole story," and Barbara agreed. The main obstacle to full disclosure disappeared with the death of Wiener's wife, Margaret, at the age of ninety-five in 1989.

The drama of Wiener's personal life begins with his years as an infant prodigy, tormented by his brilliant but tyrannical father. Either as a result of his father's training or from genetic predisposition, he suffered from violent swings of mood that continued throughout his life. If he had been seen by a modern psychiatrist he would probably have been diagnosed as manic-depressive. He sank periodically into deep depressions that continued for several months, and then emerged into intervals of restless and creative activity. The depressions tended to come more often when he stayed at home, and that was one of the reasons why he spent so much of his time traveling. Away from home, the distractions of public lectures and ceaseless conversation with friends and admirers kept his spirits high.

Another major theme of this biography is Wiener's marriage. His wife, Margaret, was a student of his father, and the marriage was arranged by his parents. Margaret was chosen to take over from his parents the job of caring for him and organizing his life. This job she performed well, running a frugal household and providing a comfortable home for him and the children. She said to a friend in the early days of the marriage, "Norbert does the math and I do the arithmetic." She coped with his moods and raised his daughters.

But Margaret was in some respects even crazier than Wiener. She had emigrated from Germany to America at the age of fourteen. She was a fervent admirer of Adolf Hitler and kept two copies of *Mein Kampf* displayed prominently in her bedroom, one in German and one in English. She made no secret of her political views, to the intense annoyance of Wiener, who was himself Jewish and had many friends who were victims of Nazi persecution. When the daughters were teenagers and began to acquire boyfriends, she made their lives miserable by accusing them of nonexistent sexual delinquencies. When they once went out with a girlfriend and came home with their ears pierced, Margaret was furious and accused them of trying to seduce their father. As a result of her paranoid accusations, both daughters escaped from home as soon as they could and thereafter had little contact with her or with Wiener.

The most tragic episode of Wiener's life happened in 1951 when he was fifty-seven years old and passionately involved in a collaboration with his friend Warren McCullough and a group of young colleagues that he called "the boys." McCullough was a neurophysiologist who had moved from Illinois to MIT to work with Wiener. They planned to explore the connections between Wiener's theory of feedback control and the functioning of living neurons and brains. "The boys" were a brilliant team, including Jerome Lettvin, who later became a leading experimental biologist. Margaret was insanely jealous of McCullough and his boys, and resolved to break up their friendship with Wiener.

At a dinner with some colleagues in Mexico, who reported the episode to Lettvin many years later, she informed Wiener that McCullough's boys had seduced his daughter Barbara when she was a teenager staying at McCullough's house.

This story had no basis in fact, but Wiener believed it. He made no attempt to verify the accusation, and immediately wrote an angry letter to the president of MIT dissolving all connection between himself and the McCullough team. From that day until the end of Wiener's life, the contact remained broken. McCullough never knew why. The effect of the breach on McCullough and his boys was devastating. The effect on Wiener was also profound. His foray into biology, and his hopes for unifying cybernetics with biology, were at an end. Margaret achieved her objective, to cut him off from his friends and have him for herself.

The personal drama of the breach between Wiener and McCullough is the centerpiece of this biography, the event around which the rest of the narrative revolves. Perhaps the authors' main purpose was to exorcise the Wiener family curse by exposing the family secrets to the light of day. Wiener is the dark hero, and Margaret is the dark villain. The book reads more like a novel than a conventional biography. And inevitably the reviewer wonders whether the story is true. Margaret is now the one who is accused and will never have a chance to answer her accusers. She never spoke with the authors, and left no friend behind to speak for her. The evidence against her is well documented and seems convincing. And still, the reviewer wonders. The evidence that Margaret claimed a seduction had taken place comes from a single informant, the late Arturo Rosenblueth, who told the story to Lettvin and others, ten years after the event. This is not the sort of evidence that would convict a murderer in a court of law. It is not likely, but possible, that Rosenblueth, who died in 1970, might have had ulterior motives for concocting the story.

This biography belongs to a genre that has recently become fashionable, emphasizing the baring of family secrets and the exposure of human weaknesses. There has been a spate of books exposing the human weaknesses of Einstein, Madame Curie, and other scientific heroes. Such books are worth reading if they give us a balanced mixture of human drama with scientific substance. Many of them make no attempt at balance, giving us stories and scandals undiluted with science. The authors of this book have succeeded in bringing Wiener to life as a great figure in the world of science as well as a tragic hero in a domestic drama. They show him as he was, a mixture of Galileo and Othello. Because they are ignorant of mathematics, they cannot give the reader a detailed picture of what Wiener actually did. But they answer the crucial questions: what cybernetics was, what Wiener intended to do with it, and why it seems to have disappeared from the scene after Wiener's death.

Wiener defined cybernetics to be "the entire field of control and communication theory, whether in the machine or in the animal." The languages of communication theory are mathematical. To understand the history of cybernetics, it is important to understand that mathematical communication has two languages, which we call analog and digital. Analog communication describes the world in terms of continuously variable quantities such as electrical voltages and currents that can be directly measured. Digital communication describes the world in terms of zeros and ones, each zero or one representing a logical choice between two discrete alternatives. Analog communication is the language of analysis. Digital communication is the language of logic.

Wiener was fluent in both languages and intended cybernetics to include both. In 1940 he wrote a memorandum explaining in detail why digital language would be preferable for the computers whose existence he already foresaw. But his own contributions to communication theory happened to be written in analog language, for four reasons. First, his work as a pure mathematician had mostly been in analysis. Second, his practical experience with antiaircraft prediction was concerned with analog measurements and analog feedback mechanisms. Third, his conversations with neurophysiologists had convinced him that the language of sensory-motor feedback signals in the brains of humans and animals is analog. Fourth, the transmission of signals by chemical hormones is evidence that the action of the brain is at least partly analog. For all these reasons, Wiener's book *Cybernetics*, which summarized his thinking in 1948, was written in analog language. And for the last ten years of his life, as he traveled from country to country preaching the gospel of cybernetics, he used analog language almost exclusively. In spite of his original intentions, cybernetics became a theory of analog processes.

Meanwhile, also in 1948, Claude Shannon published his classic pair of papers with the title "A Mathematical Theory of Communication," in *The Bell System Technical Journal*. Shannon's theory was a theory of digital communication, using many of Wiener's ideas but applying them in a new direction. Shannon's theory was mathematically elegant, clear, and easy to apply to practical problems of communication. It was far more user-friendly than cybernetics. It became the basis of a new discipline called "information theory." During the next ten years, digital computers began to operate all over the world, and analog computers rapidly became obsolete. Electronic engineers learned information theory, the gospel according to Shannon, as part of their basic training, and cybernetics was forgotten.

Neither Wiener nor von Neumann nor Shannon, nor anyone else in the 1940s, foresaw the microprocessors that would make digital computers small and cheap and reliable and available to private citizens. Nobody foresaw the Internet or the ubiquitous cell phone. As a result of the proliferation of digital computers in private hands, Wiener's nightmare vision of a few giant computers determining the fate of human societies never came to pass. But other aspects of Wiener's vision of the future are coming true. We see, as he predicted, millions of skilled human workers displaced by machines and sinking into poverty. We see the basis of the wealth of nations moving from the manufacture of goods to the processing of information. We see the beginnings of an understanding of the mysteries of the human brain. We still have much to learn from Wiener's vision.

Postscript, 2006

Each time I publish a review in *The New York Review*, I receive a bimodal set of responses. First come the responses from nonexpert readers who write to tell me how much they like the review. Second come the responses from expert readers who write to correct my mistakes. I am grateful for both categories of response, but I learn much more from the second category. It is inevitable that I make mistakes when writing about fields in which I am not an expert, and I rely on the experts to set the record straight.

This review gave me an unusually rich collection of responses in both categories. I am especially grateful to those who wrote to correct my mistakes. I have responded to their criticisms by deleting some statements and judgments that were either untrue or unfair.

23

WISE MAN

GREAT SCIENTISTS COME in two varieties, which Isaiah Berlin, quoting the seventh-century-BC poet Archilochus, called foxes and hedgehogs. Foxes know many tricks, hedgehogs only one. Foxes are interested in everything, and move easily from one problem to another. Hedgehogs are interested only in a few problems which they consider fundamental, and stick with the same problems for years or decades. Most of the great discoveries are made by hedgehogs, most of the little discoveries by foxes. Science needs both hedgehogs and foxes for its healthy growth, hedgehogs to dig deep into the nature of things, foxes to explore the complicated details of our marvelous universe. Albert Einstein was a hedgehog; Richard Feynman was a fox.

Many readers are more likely to have encountered Feynman as a storyteller, for example in his book *Surely You're Joking, Mr. Feynman!*,[1] than as a scientist. Not many are likely to have read his great textbook *The Feynman Lectures on Physics*,[2] which was a best seller among physicists but was not intended for the general public. Now we have a collection of his letters, selected and edited by his daughter, Michelle.[3]

1. Norton, 1985.

2. Addison-Wesley, 1963–1965 (three volumes).

3. *Perfectly Reasonable Deviations from the Beaten Track: The Letters of Richard P. Feynman*, edited and with an introduction by Michelle Feynman (Basic Books, 2005).

The letters do not tell us much about his science. For readers who are not scientists, it is important to understand that foxes may be as creative as hedgehogs. Feynman happened to be young at a time when there were great opportunities for foxes. The hedgehogs, Einstein and his followers at the beginning of the twentieth century, had dug deep and found new foundations for physics. When Feynman came onto the scene in the middle of the century, the foundations were firm and the universe was wide open for foxes to explore.

One of the few letters in the collection that discusses Feynman's science was written to his former student Koichi Mano. It describes the fox's way of working:

> I have worked on innumerable problems that you would call humble, but which I enjoyed and felt very good about because I sometimes could partially succeed.... The development of shock waves in explosions. The design of a neutron counter.... General theory of how to fold paper to make a certain kind of child's toy (called flexagons). The energy levels in the light nuclei. The theory of turbulence (I have spent several years on it without success). Plus all the "grander" problems of quantum theory.
>
> No problem is too small or too trivial if we can really do something about it.

"The 'grander' problems of quantum theory" were only one item in a long list of Feynman's activities.

The phrase "the 'grander' problems of quantum theory" refers to the great work for which he received a Nobel Prize in 1965: inventing the pictorial view of nature which he called "the space-time approach." This work began in 1947 as a modest enterprise, to calculate accurately the fine details of the hydrogen atom for comparison with the findings of some new experiments that had been done at Columbia University. To do the calculation, Feynman invented a new way of describing quantum processes, using pictorial diagrams instead of equations to represent interacting particles. The "Feynman diagrams" that he invented for a particular calculation caused a revolution in physics. The diagrams were not only a useful tool for calculation but a new way of understanding nature. Feynman's basic idea was simple and general. If we want to calculate a quantum process, all we need to do is to draw stylized pictures of all the interactions that can happen, calculate a number corresponding to each picture by following some simple rules, and then add the numbers together. So a quantum process is just a bundle of pictures, each of them describing a possible way in which the process can happen.

Feynman's diagrams gave us a simple visual representation of quantum processes not only for hydrogen atoms but for everything else in the universe. Within twenty years after they were invented, these diagrams became the working language of particle physicists all over the world. It is difficult now to imagine how we used to think about fields and particles before we had this language. A new book by the MIT historian David Kaiser, *Drawing Theories Apart: The Dispersion of Feynman Diagrams in Postwar Physics*,[4] gives a lively account of the spread of the diagrams, describing how they were transmitted around the world. The diagrams spread like a flu epidemic. Each new generation of young scientists became infected with the Feynman disease and then infected others with whom they came into personal contact. The Feynman epidemic lasted longer than a flu epidemic, because the incubation period was measured in years rather than in days. Many of the older scientists remained immune, but their influence waned as the new language became universal.

After Feynman's work on the diagrams was done, a year went by before it was published. He was willing and eager to share his ideas in conversation with anyone who would listen, but he found the job of

4. University of Chicago Press, 2005.

writing a formal paper distasteful and postponed it as long as he could. His seminal paper, "Space-Time Approach to Quantum Electrodynamics,"[5] might never have been written if he had not gone to Pittsburgh to stay for a few days with his friends Bert and Mulaika Corben. While he was in the Corbens' house, they urged him to sit down and write the paper, and he made all kinds of excuses to avoid doing it. Mulaika, who was a liberated woman with a forceful personality, decided that drastic measures were needed. She was one of the few people who could stand up to Feynman in a contest of wills. She locked him in his room and refused to let him out until the paper was finished. That is the story that Mulaika told me afterward. Like other Feynman stories, it may have been embellished in the telling, but to anyone who knew both Mulaika and Feynman it has the ring of truth.

People who knew Feynman as a friend and colleague were astonished when this collection of his letters appeared. We never thought of him as a letter writer. He was famous as a great scientist and a great communicator, but his way of communicating with the public was by talking rather than writing. He talked in a racy and informal style, and claimed to be incapable of writing grammatical English. His many books were not written by him but transcribed and edited by others from recordings of his talks. The technical books were records of his classroom lectures, and the popular books were records of his stories. He preferred to publish his scientific discoveries in lectures rather than in papers.

This book now reveals that Feynman was, like that other great communicator Ronald Reagan, secretly writing personal letters to a great variety of people. Few of the letters are to his professional colleagues. Many of them are to his family, and many are to people he did not know and never met, answering letters that they wrote to him with questions about science. In spite of his pretense of being

5. *Physical Review*, Vol. 76, No. 6 (September 15, 1949).

illiterate, the letters are written in lucid and grammatical English. They rarely mention his work as a creative scientist. They say nothing about his current research. In these letters we see Feynman as a teacher. He spent much of his life teaching, and he threw himself into teaching as passionately as he threw himself into research. He wrote these letters because he wanted to help anyone who sincerely tried to understand. The letters that he preferred to answer were those which posed problems that he could explain in simple language. The problems were usually elementary, and Feynman's answers were pitched at a level that his correspondent could understand. He was not trying to be clever. His purpose was to be clear.

Every one of the letters is personal. He responded to people's personal needs as well as to their questions. As an example of his personal response, here is the last paragraph of the letter to Koichi Mano which I have already quoted. Koichi was unhappy with his life as a scientist because he was not working on fundamental problems. Feynman replies:

> You say you are a nameless man. You are not to your wife and to your child. You will not long remain so to your immediate colleagues if you can answer their simple questions when they come into your office. You are not nameless to me. Do not remain nameless to yourself—it is too sad a way to be. Know your place in the world and evaluate yourself fairly, not in terms of the naïve ideals of your own youth, nor in terms of what you erroneously imagine your teacher's ideals are.
>
> Best of luck and happiness.
>
> Sincerely,
> Richard P. Feynman

Michelle Feynman added some brief comments to the letters, and an introduction describing what it was like to be Feynman's daughter.

She was as surprised as everyone else when she discovered the letters and started to read them sixteen years after his death. For sixteen years they had remained hidden in filing cabinets in the archives of the California Institute of Technology, interspersed with masses of technical papers and lecture notes. As soon as she had read them, she decided that they should be shared with the world. They show a new side of Feynman. The public had seen him before as a great scientist and as a famous clown. A week after I first met him at Cornell University in 1947, I described him in a letter to my parents as "half genius and half buffoon." Here in the letters he is neither genius nor buffoon, but a wise counselor, interested in all kinds of people, answering their questions, and trying to help them as best he can.

Michelle's introduction ends with a note that she found with the letters in the archive. Feynman wrote it for his acceptance speech at the Nobel Prize banquet in Stockholm. Before he went to Sweden, when the award of his Nobel Prize was first announced, he made disparaging remarks about the prize and about the formal ceremonies that he would have to endure in Stockholm. He said that he had made up his mind to refuse the prize, until his wife told him that refusing it would bring him even more unwelcome publicity than accepting it. He detested formal ceremonies, and he especially detested the snobbery associated with kings and queens and royal palaces. But then, after he went to Stockholm and experienced the warmth of a Swedish welcome, he wrote a note that is as close as he ever came to expressing his emotions in public. He describes how the prize had led to a deluge of messages:

> Reports of fathers turning excitedly with newspapers in hand to wives; of daughters running up and down the apartment house ringing neighbors' door bells with news; victorious cries of "I told you so" by those having no technical knowledge—their successful prediction being based on faith alone; from friends,

> from relatives, from students, from former teachers, from scientific colleagues, from total strangers....
>
> In each I saw the same two common elements. I saw in each, joy; and I saw affection (you see, whatever modesty I may have had has been completely swept away in recent days).
>
> The Prize was a signal to permit them to express, and me to learn about, their feelings....
>
> For this, I thank Alfred Nobel and the many who worked so hard to carry out his wishes in this particular way.
>
> And so, you Swedish people, with your honors, and your trumpets, and your king—forgive me. For I understand at last—such things provide entrance to the heart. Used by a wise and peaceful people they can generate good feeling, even love, among men, even in lands far beyond your own. For that lesson, I thank you.

The title of this book, *Perfectly Reasonable Deviations from the Beaten Track*, is taken from a letter that Feynman wrote for the California State Curriculum Commission, in which he appraised the science textbooks to be used in elementary schools. His son, Carl, was then three years old, due to go to elementary school three years later and learn from the textbooks. Feynman spent much time and effort reading textbooks and pointing out their deficiencies. He also examined the teachers' manuals that came with the textbooks. The manuals were supposed to explain the material in the textbooks so that teachers could teach it intelligently. Feynman was especially critical of the manuals. In response to one series of books and manuals, he wrote:

> Fair. A spotty mixture of good and poor.
>
> In 1st grade, simple clear experiments on condensation, etc., but most of the stuff on animals is how they differ in superficial ways (nothing on how they grow, eggs, babies, etc.)

> In 5th grade, chemistry and sound is good and clear, but material on weather and electricity are not very good. In particular, in both these parts (weather and electricity) the teacher's manual doesn't realize the possibilities of correct answers different from the expected ones and the teacher instruction is not enough to enable her to deal with perfectly reasonable deviations from the beaten track. Also, in these sections, difficult experiments are suggested which may not work out easily as expected, but the teacher is not given clues that this might happen or what to do about it.

He was particularly concerned that teachers using the manuals might penalize children who came up with original ways of solving problems. This actually happened many years later when Michelle was in high school and was penalized for going off the beaten track to solve an algebra problem. When Feynman went to the school to complain, the teacher accused him of knowing nothing about math. After that, Michelle stayed home to study algebra with her father, and only went to school to take exams.

Michelle was the younger of two children, doted on by her father, loving and admiring him in return. Her brother, Carl, was closer to him intellectually, sharing his interests in science and computers. Michelle describes how she tagged along silently on long walks while her father and Carl talked shop. He once complained to me about Carl:

> I always thought I was a good father, being proud of both my kids and not trying to push them into any particular direction. I did not want them to be professors like me. I would be just as happy if they were truck drivers or ballet dancers, provided they really enjoyed what they were doing. But then they always find a way to hit back at you. My son Carl for instance. He is a

> student at MIT and what does he want to do? He wants to be a God-damned philosopher.

Feynman admired people with practical skills and had no use for philosophers. Fortunately Carl's affair with philosophy was short-lived. He soon returned to computer science, a field in which he could joyfully share technical tricks and ideas with his father.

It would be easy to fill a review with quotations from the letters. At the beginning there are letters from Feynman to his parents, including a highly nontrivial arithmetical puzzle involving long division that he sent to his father when he was twenty-one years old. His father was a traveling salesman with a passion for science but without any scientific training. That puzzle must have been part of a continuing exchange of puzzles and ideas between father and son. Many years later he wrote about his father:

> He told me fascinating things about the stars, numbers, electricity.... Before I could talk he was already interesting me in mathematical designs made with blocks. So I have always been a scientist. I have always enjoyed it, and thank him for this great gift to me.

After the family letters, there is a collection of letters between Feynman and his first wife, Arline, describing day by day their doomed and difficult life through the three years between their marriage and her death from tuberculosis. For most of those three years, Feynman was working for the Manhattan Project at the atomic bomb laboratory in Los Alamos, and Arline was at a nursing home in Albuquerque, sixty miles away over rugged mountain roads. Feynman wrote to her in May 1945:

> The doc came around special to tell me of a mold growth, streptomycin, which really seems to cure TB in guinea pigs—it has

> been tried on humans—fair results except it is very dangerous as it plugs up the kidneys.... He says he thinks they may soon lick that—and if it works it will become available rapidly....
>
> Keep hanging on tho—as I say there is always a chance something will turn up. Nothing is certain. We lead a charmed life.

Streptomycin worked in humans and it became available rapidly, but not rapidly enough to save Arline. She died a month later.

Five of the letters are in a different category from the others. They were written to Feynman's third wife, Gweneth, when he was traveling, leaving her at home. Gweneth was the mother of Carl and adoptive mother of Michelle. He obviously enjoyed writing to her. The letters are full of detailed observations of things that most travelers would miss. One of his gifts was the ability to walk into a place where he had never been before and see at a glance what was going on. It is his powers of exact observation and vivid description that make the letters memorable. The same gifts made him a uniquely perceptive investigator of the *Challenger* shuttle disaster in 1986. When he was asked to serve on the commission investigating the causes of the disaster, he wanted to say no, but Gweneth said:

> If you don't do it, there will be twelve people, all in a group, going around from place to place together. But if you join the commission, there will be eleven people—all in a group, going around from place to place together—while the twelfth one runs around all over the place, checking all kinds of unusual things.... There isn't anyone else who can do that like you can.[6]

He knew Gweneth was right, and so he said yes. His "Personal

6. Richard P. Feynman, *What Do You Care What Other People Think?* (Norton, 1988), p. 117.

Observations on the Reliability of the Shuttle," published as an appendix to the official report of the commission, is full of wisdom. After looking in detail at a number of ways that accidents can happen, he concludes that the expected rate of fatal shuttle accidents is about one per hundred flights. The second fatal accident, the *Columbia* disaster of February 1, 2003, showed that his estimate was close to the truth. But the politicians and administrators who run NASA have never admitted that he was right.

The five long letters to Gweneth were written from five different places: from Brussels when he was at a physics conference in 1961 and she was pregnant with Carl, from Warsaw when he was at another physics conference in 1962, from Athens when he was lecturing there in 1980, from Switzerland when he was visiting in 1982, and from Washington when he was serving on the shuttle commission in 1986. Each of them is a set piece, a framed story of a place and time, with vivid portraits of the people that he encountered. The Brussels story is staged in the royal palace, where he was introduced to the king and queen, with a hilarious description of the stiff and stupid formalities of royal conversation. Fortunately the queen's secretary turned out to be a friendly ally, and Feynman was able to escape from the palace to spend a happy afternoon with the secretary and his wife and daughters in their country house.

The Warsaw story takes place in the dining room of the Grand Hotel, where Feynman is describing the frustrations of dealing with Communist bureaucracy. "Theoretically," he wrote,

> planning may be good etc—but nobody has ever figured out the cause of government stupidity—and until they do and find the cure all ideal plans will fall into quicksand.

The Athens story describes how the Greek educational system, with its overwhelming emphasis on the glories of classical Greece, gives children a bad start in life, teaching them that nothing they do can equal the achievements of their ancestors:

> They were very upset when I said that the thing of greatest importance to mathematics in Europe was the discovery by Tartaglia that you can solve a cubic equation—which, altho it is very little used, must have been psychologically wonderful because it showed a modern man could do something no ancient Greek could do, and therefore helped in the renaissance which was the freeing of man from the intimidation of the ancients—what they are learning in school is to be intimidated into thinking they have fallen so far below their super ancestors.

The Swiss story is the longest and the most carefully composed, with a formal title, "The Curse of Riches." It describes a visit that Feynman made to the country estate of a South American millionaire "who had inherited great wealth." He built a grandiose house similar to William Randolph Hearst's castle at San Simeon in California, with a large collection of Roman, Mayan, and Polynesian art treasures. Feynman concludes that his first favorable impressions

> had turned into a vision of surrealistic horror—as I imagined the three—he, his wife and daughter... eating alone in that long room with unpainted walls, with the Roman paintings looking down on the dark scene lit only by candles... in an ancient candlestick.... Such, in this case, is the curse of wealth.

The Washington story was written when Feynman had barely survived his second major cancer operation, less than two years before his death. It describes his continuing duel with William Rogers, the chairman of the *Challenger* commission. Feynman was determined to investigate the facts of the case wherever they might lead, and Rogers was determined to keep him on a tight leash in case he might discover something that would be politically embarrassing. Feynman treated Rogers with studied politeness and did not openly rebel. He knew that he could beat Rogers in a battle of wits. He writes to Gweneth before the battle is won, anticipating victory. He explains how Rogers hopes to keep him snowed under with data and details

> so they have time to soften up dangerous witnesses etc. But it won't work because (1) I do technical information exchange and understanding much faster than they imagine, and (2) I already smell certain rats that I will not forget because I just love the smell of rats for it is the spoor of exciting adventure.

He later wrote Rogers in defense of the commission's report:

> We have laid out the facts and done it well. The large number of negative observations are a result of the appalling condition the NASA shuttle program has gotten into. It is unfortunate, but true, and we would do a disservice if we tried to be less than frank about it.

Why should we care about Feynman? What was so special about him? Why did he become a public icon, standing with Albert Einstein and Stephen Hawking as the Holy Trinity of twentieth-century physics? The public has demonstrated remarkably good taste in choosing its icons. All three of them are genuinely great scientists, with flashes of true genius as well as solid accomplishments to their credit. But to become an icon, it is not enough to be a great scientist. There are many other scientists, not so great as Einstein but greater than Hawking and Feynman, who did not become icons. Paul Dirac is a good example of a scientist greater than Feynman. Feynman always said, whenever the opportunity arose, that the "space-time approach" that led him to his new way of doing particle physics was directly borrowed from a paper of Dirac's.[7] That was true. Dirac had the original idea and Feynman made it into a useful practical tool. Dirac was the greater genius. But Dirac did not become an icon because he had no wish to be an icon and no talent for entertaining the public.

Scientists who become icons must not only be geniuses but also performers, playing to the crowd and enjoying public acclaim. Einstein and Feynman both grumbled about the newspaper and radio reporters who invaded their privacy, but both gave the reporters what the public wanted, sharp and witty remarks that would make good headlines. Hawking in his unique way also enjoys the public adulation that his triumph over physical obstacles has earned for him. I will never forget the joyful morning in Tokyo when Hawking went on a tour of the streets in his wheelchair and the Japanese crowds streamed after him, stretching out their hands to touch his chair. Einstein, Hawking, and Feynman shared an ability to break through the barriers that separated them from ordinary people. The public responded to them because they were regular guys, jokers as well as geniuses.

The third quality that is needed for a scientist to become a public icon is wisdom. Besides being a famous joker and a famous genius, Feynman was also a wise human being whose answers to serious questions made sense. To me and to hundreds of other students who came to him for advice, he spoke truth. Like Einstein and Hawking, he had come through times of great suffering, nursing Arline through her illness and watching her die, and emerged stronger. Behind his enormous zest and enjoyment of life was an awareness of tragedy, a knowledge that our time on earth is short and precarious. The public made him into an icon because he was not only a great scientist and a

---

7. See, for example, the letter to Herbert Jehle on page 159 of *Perfectly Reasonable Deviations*.

great clown but also a great human being and a guide in time of trouble. Other Feynman books have portrayed him as a scientific wizard and as a storyteller. This collection of letters shows us for the first time the son caring for his father and mother, the husband and father caring for his wife and children, the teacher caring for his students, the writer replying to people throughout the world who wrote to him about their problems and received his full and undivided attention.[8]

8. This review sketches only one side of Feynman's many-sided personality. Other sides are sketched in Chapters 25 and 26.

IV

Personal and Philosophical Essays

24

THE WORLD, THE FLESH, AND THE DEVIL

THE WORLD, THE FLESH *and the Devil: An Enquiry into the Future of the Three Enemies of the Rational Soul* is the full title of Desmond Bernal's first book, which he published in 1929 at the age of twenty-eight.[1] Forty years later he wrote in a foreword to a reprinted edition, "This short book was the first I ever wrote. I have a great attachment to it because it contains many of the seeds of ideas which I have been elaborating throughout my scientific life. It still seems to me to have validity in its own right." It must have been a consolation to Bernal, crippled and incapacitated in the last years of his life by a stroke, to know that this work of his springtime was again being bought and read by a new generation of young readers.

The book begins: "There are two futures, the future of desire and the future of fate, and man's reason has never learnt to separate them." I do not know of any finer opening sentence of a work of literature in English. Bernal's modest claim that his book "still seems to have validity in its own right" holds good in 1972 as it did in 1968. Enormous changes have occurred since he wrote the book in 1929, both in science and in human affairs. It would be miraculous if nothing in it had become dated or superseded by the events of the last

1. New edition, Indiana University Press, 1969.

forty years. But astonishingly little of it has proved to be wrong or irrelevant to our present concerns.

Bernal saw the future as a struggle of the rational side of man's nature against three enemies. The first enemy he called the World, meaning scarcity of material goods, inadequate land, harsh climate, desert, swamp, and other physical obstacles that condemn the majority of mankind to lives of poverty. The second enemy he called the Flesh, meaning the defects in man's physiology that expose him to disease, cloud the clarity of his mind, and finally destroy him by senile deterioration. The third enemy he called the Devil, meaning the irrational forces in man's psychological nature that distort his perceptions and lead him astray with crazy hopes and fears, overriding the feeble voice of reason. Bernal had faith that the rational soul of man would ultimately prevail over these enemies. But he did not foresee cheap or easy victories. In each of the three struggles, he saw hope of defeating the enemy only if mankind is prepared to adopt extremely radical measures.

Briefly summarized, the radical measures which Bernal prescribed were the following. To defeat the World, the greater part of the human species will leave this planet and go to live in innumerable freely floating colonies scattered through outer space. To defeat the Flesh, humans will learn to replace failing organs with artificial substitutes until we become an intimate symbiosis of brain and machine. To defeat the Devil, we shall first reorganize society along scientific lines, and later learn to exercise conscious intellectual control over our moods and emotional drives, intervening directly in the affective functions of our brains with technical means yet to be discovered. This summary is a crude oversimplification of Bernal's discussion. He did not imagine that these remedies would provide a final solution to the problems of humanity. He well knew that every change in the human situation will create new problems and new enemies of the rational soul. He stopped where he stopped because he could not see any further. His chapter "The Flesh" ends with the words: "That may be an end or a beginning, but from here it is out of sight."

How much that was out of sight to Bernal in 1929 can we see from the vantage point of 1972? The first and most obvious difference between 1929 and 1972 is that we now have a highly vocal and well-organized opposition to the further growth of the part that technology plays in human affairs. The social prophets of today look upon technology as a destructive rather than a liberating force. In 1972 it is highly unfashionable to believe as Bernal did that the colonization of space, the perfection of artificial organs, and the mastery of brain physiology are the keys to man's future. People in tune with the mood of the times regard space as irrelevant, and they consider ecology to be the only branch of science that is ethically respectable. However, it would be wrong to imagine that Bernal's ideas were more in line with popular views in 1929 than they are in 1972. Bernal was never a man to swim with the tide. Technology was unpopular in 1929 because it was associated in people's minds with the gas warfare of the First World War, just as now it is unpopular by association with Hiroshima and the defoliation of Vietnam. In 1929 the dislike of technology was less noisy than today but no less real. Bernal understood that his proposals for the remaking of man and society flew in the teeth of deeply entrenched human instincts. He did not on that account weaken or compromise his statement. He believed that a rational soul would ultimately come to accept his vision of the future as reasonable, and that for him was enough. He foresaw that mankind might split into two species, one following the technological path which he described, the other holding on as best it could to the ancient folkways of natural living. And he recognized that the dispersion of mankind into the vastness of space is precisely what is required for such a split of the species to occur without intolerable strife and social disruption. The wider perspective which we have gained between 1929 and 1972 concerning the harmful effects of technology affects only the details and not the core of Bernal's argument.

Another conspicuous difference between 1929 and 1972 is that men have now visited the moon. This fact makes little difference to the plausibility of Bernal's vision. Bernal in 1929 foresaw cheap and massive emigration of human beings from the earth. He did not know how it should be done. We still do not know how it should be done. Certainly it will not be done by using the technology that took men to the moon in 1969. We know that in principle the cost in energy of transporting people from earth into space need be no greater than the cost of transporting them from New York to London. To translate this "in principle" into reality will require two things: first a great advance in the engineering of hypersonic aircraft, and second the growth of a traffic massive enough to permit large economies of scale. It is likely that the *Apollo* vehicle bears the same relation to the cheap mass-transportation space vehicle of the future as the majestic airship of the 1930s bears to the Boeing 747 of today. The airship *R101* was absurdly large, beautiful, expensive, and fragile, just like the *Apollo Saturn 5*. If this analogy is sound, we shall have transportation into space at a reasonable price within about fifty years from now. But my grounds for believing this are not essentially firmer than Bernal's were for believing it in 1929.

The decisive change that has enabled us to see farther in 1972 than Bernal could see in 1929 is the advent of molecular biology. Bernal was himself one of the founding fathers of molecular biology. In the 1930s he mastered the art of mapping the structure of large molecules by means of X-rays. He understood that this art would be the key to the understanding of the physical basis of life. His pioneering work led directly to the discovery of the double helix in 1953. Rosalind Franklin, who took the crucial X-ray pictures of DNA that showed the helical structure, was working in Bernal's laboratory in London. In the 1968 foreword to his book, Bernal speaks of the double helix as "the greatest and most comprehensive idea in all science." As a result of this discovery, we understand the basic principles by which living cells organize and reproduce themselves. Many mysteries remain, but it is inevitable that we shall understand the chemical processes of life in full detail, including the processes of development and differentiation of higher organisms, within the next century. I consider it also inevitable and desirable that we shall learn to exploit these processes for our own purposes. The next century will see a completely new technology growing out of the mastery of the principles of biology, just as our existing technology grew out of a mastery of the principles of physics.

The new biological technology may grow in three distinct directions. Probably all three will be followed and will prove fruitful for particular purposes. The first direction is the one that has been chiefly discussed by biologists who feel responsibility for the human consequences of their work; they call it "genetic surgery." The idea is that we shall be able to read the base sequence of the DNA in a human sperm or egg cell, run the sequence through a computer which will identify deleterious genes or mutations, and then by micromanipulation patch harmless genes into the sequence to replace the bad ones. It might also be possible to add to the DNA genes conferring various desired characteristics to the resulting individual. This technology will be difficult and dangerous, and its use will raise severe ethical problems. Jacques Monod in his 1971 book *Chance and Necessity* sweeps all thought of it aside with his customary dogmatic certitude. "There are," he says, "occasional promises of remedies expected from the current advances in molecular genetics. This illusion, spread about by a few superficial minds, had better be disposed of." Although I have a great respect for Monod, I still dare to brave his scorn by stating my belief that genetic surgery has an important part to play in man's future. But I share the prevailing view of biologists that we must be exceedingly careful in interfering with the human genetic material. The interactions between the thousands of genes in a human cell are so exquisitely complicated that a computer program labeling genes "good" or "bad" will be adequate to deal only with the grossest sort of defect. There are strong arguments for declaring a moratorium on genetic surgery for the next hundred years, or until we understand human genetics vastly better than we do now.

Leaving aside genetic surgery applied to humans, I foresee that the coming century will place in our hands two other forms of biological technology which are less dangerous but still revolutionary enough to transform the conditions of our existence. I count these new technologies as powerful allies in the attack on Bernal's three enemies. I give them the names "biological engineering" and "self-reproducing machinery." Biological engineering means the artificial synthesis of living organisms designed to fulfill human purposes. Self-reproducing machinery means the imitation of the function and reproduction of a living organism with nonliving materials, a computer program imitating the function of DNA, and a miniature factory imitating the functions of protein molecules. After we have attained a complete understanding of the principles of organization and development of a simple multicellular organism, both of these avenues of technological exploitation should be open to us.

I would expect the earliest and least controversial triumphs of biological engineering to be extensions of the art of industrial fermentation. When we are able to produce microorganisms equipped with enzyme systems tailored to our own design, we can use such organisms to perform chemical operations with far greater delicacy and economy than present industrial practices allow. For example, oil refineries would contain a variety of bugs designed to metabolize crude petroleum into the precise hydrocarbon isomers which are needed for various purposes. One tank would contain the n-octane bug, another the benzene bug, and so on. All the bugs would contain enzymes metabolizing sulfur into elemental form, so that pollution of the atmosphere by sulfurous gases would be completely controlled. The management and operation of such fermentation tanks on a vast scale would not be easy, but the economic and social rewards are so great that I am confident we shall learn how to do it. After we have mastered the biological oil refinery, more important applications of the same principles will follow. We shall have factories producing specific foodstuffs biologically from cheap raw materials, and sewage treatment plants converting our wastes efficiently into usable solids and pure water. To perform these operations we shall need an armamentarium of many species of microorganisms trained to ingest and excrete the appropriate chemicals. And we shall design into the metabolism of these organisms the essential property of self-liquidation, so that when deprived of food they disappear by cannibalizing one another. They will not, like the bacteria that feed upon our sewage in today's technology, leave their rotting carcasses behind to make a sludge only slightly less noxious than the mess that they have eaten.

If these expectations are fulfilled, the advent of biological technology will help enormously in the establishment of patterns of industrial development with which human beings can live in health and comfort. Oil refineries need not stink. Rivers need not be sewers. However, there are many environmental problems which the use of artificial organisms in enclosed tanks will not touch. For example, the fouling of the environment by mining and by abandoned automobiles will not be reduced by building cleaner factories. The second step in biological engineering, after the enclosed biological factory, is to let artificial organisms loose into the environment. This is admittedly a more dangerous and problematical step than the first. The second step should be taken only when we have a deep understanding of its ecological consequences. Nevertheless the advantages which artificial organisms offer in the environmental domain are so great that we are unlikely to forgo their use forever.

The two great functions which artificial organisms promise to perform for us when let loose upon the earth are mining and scavenging.

The beauty of a natural landscape undisturbed by man is largely due to the fact that the natural organisms in a balanced ecology are excellent miners and scavengers. Mining is mostly done by plants and microorganisms extracting minerals from water, air, and soil. For example, it has been recently discovered that organisms in the ground mine ammonia and carbon monoxide from air with high efficiency. To the scavengers we owe the fact that a natural forest is not piled as high with dead birds as one of our junkyards with dead cars. Many of the worst offenses of human beings against natural beauty are due to our incompetence in mining and scavenging. Natural organisms know how to mine and scavenge effectively in a natural environment. In a man-made environment, neither they nor we know how to do it. But there is no reason why we should not be able to design artificial organisms that are adaptable enough to collect our raw materials and to dispose of our refuse in an environment that is a careful mixture of natural and artificial.

A simple example of a problem that an artificial organism could solve is the eutrophication of lakes. At present many lakes are being ruined by excessive growth of algae feeding on high levels of nitrogen or phosphorus in the water. The damage could be stopped by an organism that would convert nitrogen to molecular form or phosphorus to an insoluble solid. Alternatively and preferably, an organism could be designed to divert the nitrogen and phosphorus into a food chain culminating in some species of palatable fish. To control and harvest the mineral resources of the lake in this way will in the long run be more feasible than to maintain artificially a state of "natural" barrenness.

The artificial mining organisms would not operate in the style of human miners. Many of them would be designed to mine the ocean. For example, oysters might extract gold from seawater and secrete golden pearls. A less poetic but more practical possibility is the artificial coral that builds a reef rich in copper or magnesium. Other mining organisms would burrow like earthworms into mud and clay, concentrating in their bodies the ores of aluminum or tin or iron, and excreting the ores in some manner convenient for human harvesting. Almost every raw material necessary for our existence can be mined from ocean, air, or clay, without digging deep into the earth. Where conventional mining is necessary, artificial organisms can still be useful for digesting and purifying the ore.

Not much imagination is needed to foresee the effectiveness of artificial organisms as scavengers. A suitable microorganism could convert the dangerous organic mercury in our rivers and lakes to a harmless insoluble solid. We could make good use of an organism with a consuming appetite for polyvinyl chloride and similar plastic materials which now litter beaches all over the Earth. Conceivably we may produce an animal specifically designed for chewing up dead automobiles. But one may hope that the automobile in its present form will become extinct before it needs to be incorporated into an artificial food chain. A more serious and permanent role for scavenging organisms is the removal of trace quantities of radioactivity from the environment. The three most hazardous radioactive elements produced in fission reactors are strontium, caesium, and plutonium. These elements have long half-lives and will inevitably be released in small quantities so long as mankind uses nuclear fission as an energy source. The long-term hazard of nuclear energy would be notably reduced if we had organisms designed to gobble up these three elements from water or soil and to convert them into indigestible form. Fortunately, none of these three elements is essential to our body chemistry, and it therefore does us no harm if they are made indigestible.

I have described the two first steps of biological engineering. The first will transform our industry and the second will transform our earthbound ecology. It is now time to describe the third step, which is the colonization of space. Biological engineering is the essential tool which will make Bernal's dream of the expansion of mankind in space a practical possibility.

First I have to clear away a few popular misconceptions about space as a habitat. It is generally considered that planets are important. Except for Earth, they are not. Mars is waterless, and the others are for various reasons basically inhospitable to man. It is generally considered that beyond the sun's family of planets there is absolute emptiness extending for light-years until you come to another star. In fact it is likely that the space around the solar system is populated by huge numbers of comets, small worlds a few miles in diameter, rich in water and the other chemicals essential to life. We see one of these comets only when it happens to suffer a random perturbation of its orbit which sends it plunging close to the sun. It seems that roughly one comet per year is captured into the region near the sun where it eventually evaporates and disintegrates. If we assume that the supply of distant comets is sufficient to sustain this process over the billions of years that the solar system has existed, then the total population of comets loosely attached to the sun must be numbered in the billions. The combined surface area of these comets is then a thousand or ten thousand times that of Earth. I conclude from these facts that comets, not planets, are the major potential habitat of life in space. If it were true that other stars have as many comets as the sun, it then would follow that comets pervade our entire galaxy. We have no evidence either supporting or contradicting this hypothesis. If true, it implies that our galaxy is a much friendlier place for interstellar travelers than it is popularly supposed to be. The average distance between habitable oases in the desert of space is not measured in light-years, but is of the order of a light-day or less.

I propose then an optimistic view of the galaxy as an abode of life. Countless millions of comets are out there, amply supplied with water, carbon, and nitrogen, the basic constituents of living cells. We see when they fall close to the sun that they contain all the common elements necessary to our existence. They lack only two essential requirements for human settlement, namely warmth and air. And now biological engineering will come to our rescue. We shall learn to grow trees on comets.

To make a tree grow in airless space by the light of a distant sun is basically a problem of redesigning the skin of its leaves. In every organism the skin is the crucial part which must be most delicately tailored to the demands of the environment. The skin of a leaf in space must satisfy four requirements. It must be opaque to far-ultraviolet radiation to protect the vital tissues from radiation damage. It must be impervious to water. It must transmit visible light to the organs of photosynthesis. It must have extremely low emissivity for far-infrared radiation, so that it can limit loss of heat and keep itself from freezing. A tree whose leaves possess such a skin should be able to take root and flourish upon any comet as near to the sun as the orbits of Jupiter and Saturn. Further out than Saturn the sunlight is too feeble to keep a simple leaf warm, but trees can grow at far greater distances if they provide themselves with compound leaves. A compound leaf would consist of a photosynthetic part which is able to keep itself warm, together with a concave mirror part which itself remains cold but focuses concentrated sunlight upon the photosynthetic part. It should be possible to program the genetic instructions of a tree to produce such leaves and orient them correctly toward the sun. Many existing plants possess structures more complicated than this.

Once leaves can be made to function in space, the remaining parts of a tree—trunk, branches, and roots—do not present any great problems. The branches must not freeze, and therefore the bark must be a superior heat insulator. The roots will penetrate and gradually melt the frozen interior of the comet, and the tree will build its substance from the materials which the roots find there. The oxygen which the leaves manufacture must not be exhaled into space. Instead it will be transported down to the roots and released into the regions where humans will live and take their ease among the tree trunks. One question still remains. How high can a tree on a comet grow? The answer is surprising. On any celestial body whose diameter is of the order of ten miles or less, the force of gravity is so weak that a tree can grow infinitely high. Ordinary wood is strong enough to lift its own weight to an arbitrary distance from the center of gravity. This means that from a comet of ten-mile diameter trees can grow out for hundreds of miles, collecting the energy of sunlight from an area thousands of times larger than the area of the comet itself. Seen from far away, the comet will look like a small potato sprouting an immense growth of stems and foliage. When humans come to live on the comets, they will find themselves returning to the arboreal existence of their ancestors.

We shall bring to the comets not only trees but a great variety of other flora and fauna to create for ourselves an environment as beautiful as ever existed on Earth. Perhaps we shall teach our plants to make seeds which will sail out across the ocean of space to propagate life upon comets still unvisited by humans. Perhaps we shall start a wave of life which will spread from comet to comet without end until we have have achieved the greening of the galaxy. That may be an end or a beginning, as Bernal said, but from here it is out of sight.

In parallel with our exploitation of biological engineering, we may achieve an equally profound industrial revolution by following the alternative route of self-reproducing machinery. Self-reproducing machines are devices which have the multiplying and self-organizing capabilities of living organisms but are built of metal and computers instead of protoplasm and brains. It was the mathematician John von Neumann who first demonstrated that self-reproducing machines are theoretically possible and sketched the logical principles underlying their construction. The basic components of a self-reproducing machine are precisely analogous to those of a living cell. The separation of function between genetic material (DNA) and enzymatic machinery (protein) in a cell corresponds exactly to the separation between software (computer programs) and hardware (machine tools) in a self-reproducing machine.

I assume that in the next century, partly imitating the processes of life and partly improving on them, we shall learn to build self-reproducing machines programmed to multiply, differentiate, and coordinate their activities as skillfully as the cells of a higher organism such as a bird. After we have constructed a single egg-machine and supplied it with the appropriate computer program, the egg and its progeny will grow into an industrial complex capable of performing economic tasks of arbitrary magnitude. It can build cities, plant gardens, construct electric power-generating facilities, launch spaceships, or raise chickens. The overall programs and their execution will remain always under human control.

The effects of such a powerful and versatile technology on human affairs are not easy to foresee. Used unwisely, it offers a rapid road to ecological disaster. Used wisely, it offers a rapid alleviation of all the purely economic difficulties of mankind. It offers to rich and poor nations alike a rate of growth of economic resources so rapid that economic constraints will no longer be dominant in determining how people are to live. In some sense this technology will constitute a permanent solution of mankind's economic problems. Just as in the past, when economic problems cease to be pressing, we shall find no lack of fresh problems to take their place.

It may well happen that on Earth, for aesthetic or ecological reasons, the use of self-reproducing machines will be strictly limited and the methods of biological engineering will be used instead wherever this alternative is feasible. For example, self-reproducing machines could proliferate in the oceans and collect minerals for human use, but we might prefer to have the same job done more quietly by corals and oysters. If economic needs were no longer paramount, we could afford a certain loss of efficiency for the sake of a harmonious environment. Self-reproducing machines may therefore play on Earth a subdued and self-effacing role.

The true realm of self-reproducing machinery will be in those regions of the solar system that are inhospitable to humans. Machines built of iron, aluminum, and silicon have no need of water. They can flourish and proliferate on the moon or on Mars or among the asteroids, carrying out gigantic industrial projects at no risk to the Earth's ecology. They will feed upon sunlight and rock, needing no other raw material for their construction. They will build in space the freely floating cities that Bernal imagined for human habitation. They will bring oceans of water from the satellites of the outer planets, where it is to be had in abundance, to the inner parts of the solar system where it is needed. Ultimately this water will make even the deserts of Mars bloom, and humans will walk there under the open sky breathing air like the air of Earth.

Taking a long view into the future, I foresee a division of the solar system into two domains. The inner domain, where sunlight is abundant and water scarce, will be the domain of great machines and governmental enterprises. Here self-reproducing machines will be obedient slaves, and people will be organized in giant bureaucracies. Outside and beyond the sunlit zone will be the outer domain, where water is abundant and sunlight scarce. In the outer domain lie the comets where trees and humans will live in smaller communities, isolated from one another by huge distances. Here humans will find once again the wilderness that they have lost on Earth. Groups of people will be free to live as they please, independent of governmental authorities. Outside and away from the sun, they will be able to wander forever on the open frontier that this planet no longer possesses.

I have described how we may deal with the World and the Flesh, and I have said nothing about how we may deal with the Devil. Bernal also had difficulties with the Devil. He admitted in the 1968 foreword to his book that the chapter on the Devil was the least satisfactory part of it. The Devil will always find new varieties of human folly to frustrate our too rational dreams.

Instead of pretending that I have an antidote to the Devil's wiles, I end with a discussion of the human factors that most obviously stand in the way of our achieving the grand designs which I have been describing. When mankind is faced with an opportunity to embark on any great undertaking, there are always three human weaknesses that devilishly hamper our efforts. The first is an inability to define or agree upon our objectives. The second is an inability to raise sufficient funds. The third is the fear of a disastrous failure. All three factors have been conspicuously plaguing the United States space program in recent years. It is a remarkable testimony to the vitality of the program that these factors still have not succeeded in bringing it to a halt. When we stand before the far greater enterprises of biological technology and space colonization that lie in our future, the same three factors will certainly rise again to confuse and delay us.

I want now to demonstrate to you by a historical example how these human weaknesses may be overcome. I shall quote from William Bradford, one of the Pilgrim Fathers, who wrote a book, *Of Plimoth Plantation*, describing the history of the first English settlement in Massachusetts. Bradford was governor of the Plymouth colony for twenty-eight years. He began to write his history ten years after the settlement. His purpose in writing it was, as he said, "That their children may see with what difficulties their fathers wrestled in going through these things in their first beginnings. As also that some use may be made hereof in after times by others in such like weighty employments." Bradford's work remained unpublished for two hundred years, but he never doubted that he was writing for the ages.

Here is Bradford describing the problem of inability to agree upon objectives. The date is spring 1620, the same year in which the Pilgrims were to sail:

> But as in all businesses the acting part is most difficult, especially where the work of many agents must concur, so was it found in this. For some of those that should have gone in England fell off and would not go; other merchants and friends that

> had offered to adventure their moneys withdrew and pretended many excuses; some disliking they went not to Guiana; others again would adventure nothing except they went to Virginia. Some again (and those that were most relied on) fell in utter dislike with Virginia and would do nothing if they went thither. In the midst of these distractions, they of Leyden who had put off their estates and laid out their moneys were brought into a great strait, fearing what issue these things would come to.

The next quotation deals with the perennial problem of funding. Here Bradford is quoting a letter written by Robert Cushman, the man responsible for buying provisions for the Pilgrims' voyage. He writes from Dartmouth on August 17, 1620, desperately late in the year, months after the ships ought to have started:

> And Mr. Martin, he said he never received no money on those conditions; he was not beholden to the merchants for a pin, they were bloodsuckers, and I know not what. Simple man, he indeed never made any conditions with the merchants, nor ever spake with them. But did all that money fly to Hampton, or was it his own? Who will go and lay out that money so rashly and lavishly as he did, and never know how he comes by it or on what conditions? Secondly, I told him of the alteration long ago and he was content, but now he domineers and said I had betrayed them into the hands of slaves; he is not beholden to them, he can set out two ships himself to a voyage. When, good man? He hath but fifty pounds in and if he should give up his accounts he would not have a penny left him, as I am persuaded. Friend, if ever we make a plantation, God works a miracle, especially considering how scant we shall be of victuals, and most of all ununited amongst ourselves and devoid of good tutors and regiment.

My last quotation describes the fear of disaster, as it appeared in the debate among the Pilgrims over their original decision to go to America:

> Others again, out of their fears, objected against it and sought to divert from it; alleging many things, and those neither unreasonable nor improbable; as that it was a great design and subject to many inconceivable perils and dangers; as, besides the casualties of the sea (which none can be freed from), the length of the voyage was such as the weak bodies of women and other persons worn out with age and travail (as many of them were) could never be able to endure. And yet if they should, the miseries of the land which they should be exposed unto, would be too hard to be borne and likely, some or all of them together, to consume and utterly to ruinate them. For there they should be liable to famine and nakedness and the want, in a manner, of all things. The change of air, diet, and drinking of water would infect their bodies with sore sicknesses and grievous diseases. And also those which should escape or overcome these difficulties should yet be in continual danger of the savage people, who are cruel, barbarous and most treacherous, being most furious in their rage and merciless where they overcome; not being content only to kill and take away life, but delight to torment men in the most bloody manner that may be.

I could go on quoting Bradford for hours, but this is not the place to do so. What can we learn from him? We learn that the three devils of disunity, shortage of funds, and fear of the unknown are no strangers to humanity. They have always been with us and will always be with us, whenever great adventures are contemplated. From Bradford we learn too how they are to be defeated. The Pilgrims used no technological magic to defeat them. The Pilgrims' victory demanded the full range of virtues of which human beings under stress are capable;

toughness, courage, unselfishness, foresight, common sense, and good humor. Bradford would have set at the head of this list the virtue he considered most important, a faith in Divine Providence.

I end this sermon on a note of disagreement with Bernal. He believed that we shall defeat the Devil by means of a combination of socialist organization and applied psychology. I believe that our best defense will be to rely on the human qualities that have remained unchanged from Bradford's time to ours. If we are wise, we shall preserve intact these qualities of the human species through the centuries to come, and they will see us safely through the many crises of destiny that surely await us. But I will let Bernal have the last word. Bernal's last word is a question which Bradford must often have pondered, but would not have known how to answer, as he watched the first generation of native-born New Englanders depart from the ways of their fathers:

> We hold the future still timidly, but perceive it for the first time, as a function of our own action. Having seen it, are we to turn away from something that offends the very nature of our earliest desires, or is the recognition of our new powers sufficient to change those desires into the service of the future which they will have to bring about?[2]

---

2. This lecture was given at Birkbeck College in London, originally founded as a night school where working people with an ambition to better themselves could get an education. Bernal, as a radical Marxist, found Birkbeck congenial and spent most of his working life there. He died in 1971 and I gave the lecture in his memory a year later. The ideas expressed in the lecture provided the foundation for most of my later writings about the future, especially the chapters "Thought-experiments" and "The Greening of the Galaxy" in my book *Disturbing the Universe* (Harper and Row, 1979).

25

IS GOD IN THE LAB?

HERE ARE TWO famous scientists expressing their opinions about science and religion. Richard Feynman gave a series of Danz Lectures at the University of Washington in Seattle in 1963.[1] John Polkinghorne gave a series of Terry Lectures at Yale University in 1996.[2] Two characters as different as it is possible to be, Polkinghorne the conscientious academic scholar, Feynman the impulsive rebel. Polkinghorne prepared the text of his lectures carefully for publication, giving us a polished and logically coherent argument. Feynman was invited to prepare a text for the University of Washington Press to publish in 1963, but never did. The University of Washington recorded the lectures and preserved the tapes.

What we have here is a verbatim transcript of the lectures as Feynman gave them, speaking extemporaneously from fragmentary notes. Feynman's voice and personality come through clearly. He talks about real people and their problems, not about philosophical abstractions. He is interested in religion as a way for people to make sense of their lives, but he is not interested in theology. Polkinghorne has the opposite bias. He is a scientist who is also an ordained minister of the

1. *The Meaning of It All: Thoughts of a Citizen-Scientist* (Addison-Wesley, 1998).

2. *Belief in God in an Age of Science* (Yale University Press, 1998).

Church of England. To be ordained, he went through formal training in theology. For him, theology is as real and as serious as science. His book has more to say about theology than about religion.

To display the contrasting styles of the two books, I pick out an outstanding passage from each. From Polkinghorne I pick out his second chapter, with the title "Finding Truth: Science and Religion Compared." This is a remarkable tour de force. Polkinghorne compares two historic intellectual struggles, one from science and one from religion. From science he takes the discovery and development of quantum mechanics, a struggle that has lasted from the beginning to the end of the twentieth century. From religion he takes the theological understanding of the nature of Jesus, a struggle that lasted from the time when Saint Paul was writing his letters shortly after Jesus' death to the modern era of diverse views and diminished certainties. He divides each of the two struggles into five periods, and shows how events in each of the five periods in the development of quantum mechanics correspond in detail to events in the matching period in the development of theology. In the first period, the breakdown of classical mechanics, the enigma of atomic spectra, and the discovery of the light-quantum by Max Planck and Albert Einstein correspond to the death of Jesus, the enigma of his resurrection as experienced by his disciples in Jerusalem, and the new understanding of these events by Saint Paul.

In the second period, confusion reigns both in physics and in theology: classical and quantum pictures in conflict in physics, orthodoxies and heresies in conflict in theology. In the third period, there was the great triumph of quantum mechanics as it emerged in 1925 and solved most of the outstanding problems of physics, and the great triumph of Christology in the year 451, when the assembled theologians at the Council of Chalcedon promulgated the doctrine concerning the nature of Jesus that orthodox Christians were thereafter required to believe. In the fourth period, a continued wrestling with unsolved problems, the paradoxes of interpretation of quantum theory in physics and the paradoxes of the incarnation of Jesus in theology. In the fifth period, recognition in both physics and theology that the new insights have deep implications and that we are very far from any final truth.

Polkinghorne argues from the detailed concordance of the two struggles that science and theology are two aspects of a single intellectual adventure. He sees theology as dealing with God in essentially the same way as science is dealing with nature. This is a grand vision. The historical evidence that he brings to support it is impressive. But I have to say that, much as I admire Polkinghorne's vision, I cannot share it. To share it, you must disregard a crucial difference between science and theology. When all is said and done, science is about things and theology is about words. Things behave in the same way everywhere, but words do not. Quantum mechanics works equally in all countries and in all cultures. Quantum mechanics gives plants the power to turn the energy of sunlight into leaves and fruit, and it gives animals the power to turn the energy of sunlight into neural images in retinas and brains, whether they are living in Tokyo or in Timbuktu. Theology works in one culture alone. If you have not grown up in Polkinghorne's culture, where words such as "incarnation" and "trinity" have a profound meaning, you cannot share his vision.

A striking passage of Feynman's book runs from page 34 to page 48. Like the Polkinghorne passage, it comes in the second chapter and deals explicitly with the relation between science and religion. But there the similarity of the two passages ends. Feynman has no interest in scholastic arguments. He is concerned only with human problems. He has a deep respect for religion, because he sees it as helping people to behave well toward one another and to be brave in facing tragedy. He respects religion as an important part of human nature. He does not himself believe in God, but he has no wish to destroy other people's belief. He does not write about professional scientists or professional religious thinkers. He writes about students who come to college from homes where religious belief is strong, and then find that exposure to modern science is calling their beliefs into question. He has seen at first hand the anguish that some of these students experience. He does not claim to have a cure for their anguish. He sees a genuine conflict between the old-fashioned family religion that commands the students to believe without question, and the ethic of science that commands them to question everything.

The conflict is acute for students who have grown up in fundamentalist Christian households. For them, the only ways out of their dilemma may be to reject science altogether or to abandon their religious heritage. Fortunately, they are a small minority of Christians. Most Christian believers are able to reconcile their general belief in God and in the teachings of Jesus with a considerable skepticism about details. For the majority, religion is a way to live rather than a set of dogmatic beliefs. Just as science can live without certainty, religion can live without dogma, and the two can live together without conflict. This is the solution that Feynman recommends to his students.

But Feynman cannot remain solemn for long. Scattered through his book is a wonderful collection of personal stories, told in the authentic Feynman style, bringing his meditations to life. One can see that for him the stories are more important than the philosophy. Most of the stories are funny; a few of them are sad. He knows how to make even the deepest personal tragedy into a good story. The most memorable of the stories is told as an example of a fake miracle. This happened at the worst moment of his life, when his first wife died of tuberculosis, after a long illness, at 9:22 in the evening. The clock in her room stopped at 9:22 and never ran again. If this had happened to a religious person, it might well have been described as a miracle. But Feynman, with his skeptical spirit, was watching at that moment and saw what really happened. The nurse who was also in the room had to fill out the death certificate to record the time and manner of his wife's death. Since the light in the room was dim, the nurse picked up the clock to read the time. She read it and then put it down. The clock was old and worn-out and had a habit of stopping if it was disturbed. So that was how the miracle happened, and perhaps it is also the way other miracles happen. For Feynman, religion has more to do with psychology than with theology. Religion is for Feynman an important part of human nature, to be examined like other human phenomena with the skeptical eye of science.

It is a curious accident of history that the Christian religion became heavily involved with theology. No other religion finds it necessary to formulate elaborately precise statements about the abstract qualities and relationships of gods and humans. There is nothing analogous to theology in Judaism or in Islam. I do not know much about Hinduism and Buddhism, but my Asian friends tell me that these religions also have no theology. They have beliefs and stories and ceremonies and rules of behavior, but their literature is poetic rather than analytical. The idea that God may be approached and understood through intellectual analysis is uniquely Christian.

The prominence of theology in the Christian world has had two important consequences for the history of science. On the one hand, Western science grew out of Christian theology. It is probably not an accident that modern science grew explosively in Christian Europe and left the rest of the world behind. A thousand years of theological disputes nurtured the habit of analytical thinking that could also be applied to the analysis of natural phenomena. On the other hand, the close historical relations between theology and science have caused conflicts between science and Christianity that do not exist between science and other religions. It is more difficult for a modern scientist to be a serious Christian, like Polkinghorne, than to be a serious Muslim, like the Nobel Prize–winning physicist Abdus Salam. Salam happily proclaimed his Muslim faith but did not feel any need to write books about it. For Salam, the idea of a conflict between his faith and his science was ludicrous. Muslim faith has nothing to do with science. But Polkinghorne writes books to prove to himself and to us that his theology and his science can live together harmoniously. For him the possibility of conflict is real, because his theology and his science sprang from the same root.

The common root of modern science and Christian theology was Greek philosophy. The historical accident that caused the Christian religion to become heavily theological was the fact that Jesus was born in the eastern part of the Roman Empire at a time when the prevailing culture was profoundly Greek.

I had the good luck a few years ago to visit the archaeological site of Zippori in Israel, with one of the Israeli archaeologists who are excavating the city as my guide. Zippori is the Hebrew name for the city. The Romans called it Sepphoris, which is its Greek name. This visit made me acutely aware of the paradox at the heart of the Christian religion. I could see here displayed the Greek culture that Jesus decisively rejected, the same Greek culture that infiltrated the Christian religion very soon after his death and has dominated Christianity ever since. Zippori is very close to where Jesus lived as a boy and as a young man. I climbed to the top of the highest building in Zippori and looked across to the next hill five miles away. On the next hill is Nazareth. Today, Nazareth is a city and Zippori is a ruin. In the time of Jesus, Zippori was a city and Nazareth was a village. From Nazareth to Zippori is an easy walk.

Two features of Zippori strike the eye immediately. First, the mikveh, or ritual baths, dug deep in the ground under every house, still survive. They prove that the inhabitants were pious Jews. Second, the mosaic decorations, some of the finest in the world, are Greek in style and subject matter. One particularly well-preserved mosaic shows the Greek hero Heracles winning a drinking contest. Some of the mosaics contain inscriptions, all in Greek. They prove that the inhabitants were thoroughly Hellenized. The inhabitants lived well, in a world where Greek was the language of wealth and education,

Hebrew the language of religion, Aramaic the language of peasants. Jesus, as we know, spoke Aramaic. The gospels describing his life, and the letters of Paul defining the new religion, were written in Greek.

After examining the evidence of the lifestyle of Zippori, one is not surprised to learn that in the great Jewish revolt against Rome in 70 CE, in which Jerusalem was destroyed, Zippori sided with Rome and escaped destruction. In the second Jewish revolt when Hadrian was emperor, Zippori sided with Rome and again escaped. A little later, the chief rabbi of Zippori was a personal friend of the Roman emperor Caracalla. No doubt they talked in Greek when they discussed the affairs of the empire and of its Jewish citizens. Zippori was finally destroyed, not by war but by an earthquake, after the empire became officially Christian. Archaeologists love earthquakes, because they leave the patterns of daily life intact under the rubble. After the earthquake, nobody came back to live at Zippori. Nobody disturbed the ruins.

Local legend in the Christian Arab community says that Jesus' mother was born and grew up in Zippori. The legend is consistent with the few facts we know about Mary. She may well have been a city girl who moved out to the village of Nazareth after she was betrothed to Joseph. Whether the legend is true or not, it is likely that Joseph walked or rode to the city from time to time, buying provisions in the market and perhaps also selling his carpenter's wares. And as soon as Jesus was old enough to walk five miles, he must have walked over to the city too. It is impossible to imagine a brilliant child growing up within five miles of a major city and not taking every chance to explore it. Zippori was at that time one of the two largest cities in the province of Galilee, the other being Tiberias.

The Bible story tells us that when Jesus was twelve years old and his family was visiting Jerusalem, he seized the opportunity to spend three days talking with the learned doctors in the Temple, and all that heard him were astonished at his understanding. Whether that story is true or not, he must have had many opportunities to sharpen his understanding by talking with the learned doctors in Zippori. I imagine that Zippori was the place where he got to know the scribes and Pharisees that he afterward so vehemently denounced. It is likely that as an adolescent he was deeply immersed in the Greek culture of the city. It is certain that as an adult he reacted violently against it. Although Zippori must have been an important part of his life, it is not once mentioned in the Bible.

I also visited Kefar Nahum, the site on the shore of the lake of Galilee that is called Capernaum in the Bible. From the biblical account one has the impression that Capernaum was a fishing village and that the disciples Peter and Andrew were simple fishermen when Jesus found them there. The real Capernaum was a Greek city, not as large as Zippori but sharing the same style and culture. There is a well-preserved synagogue that looks like a Greek temple. The city was spacious in the Greek style. It had large public buildings, and open spaces where one can imagine young men hanging out and discussing the latest ideas in philosophy and religion. After visiting Capernaum, I no longer think of Peter and Andrew as simple fishermen. I think of them as young men about town, who made a living by fishing but were also immersed in the Greek culture of the city. When Jesus came down from the hills and called them to leave their homes and share with him the rugged life of an itinerant preacher, they knew what they were leaving and why. Probably they already shared Jesus' hatred for the hypocrisies of city life, and that was why they came when he called them.

I am painting a romantic picture of Jesus and his disciples, based on fragmentary archaeological evidence. Whether this picture is true or not, two facts are certain. First, Jesus was no simple peasant, but grew up in intimate contact with an urban and overwhelmingly Greek culture. And second, he intended to lead a spiritual regeneration of his people, based on a total repudiation of Greek culture. In all his preaching, he quotes from the Law and the Prophets, the old

Hebrew scriptures. After seeing what the Greek culture had to offer, he went back to his Hebrew roots.

When Jesus died, he left behind a mass movement that rapidly grew into a new religion. The new religion moved fast from Jerusalem to other cities that were easily accessible to travelers, cities where Greek culture was even more predominant. The followers of Jesus first called themselves Christians in the Greek city of Antioch. And the man who took charge of the new religion, Saint Paul of Tarsus, was a thoroughly Hellenized Jew. Saint Paul preached to the learned men of Athens in their own language. In his writings he laid the foundations for what became orthodox Christian doctrine. Christianity became a religion for people ignorant of Hebrew and educated in the Greek tradition. Within a century, the Greek culture had swept over Christianity, and Greek philosophy had metamorphosed into Christian theology.

This history has left Western civilization with a strangely divided legacy. On the one hand, the religion of Jesus as we find it in his teachings recorded in the gospels, a religion for ordinary people trying to find their way in a harsh world. On the other hand, the theology that turned the Christian religion into a demanding intellectual discipline, a breeding ground for scholars and ultimately for scientists. Feynman is writing about the first, Polkinghorne about the second. There is not much connection between them. It is one of the great ironies of history that Jesus gave rise to them both.

Polkinghorne ends his book with a quiet statement of the faith that the scientist and the theologian hold in common, the faith that the understanding of God or nature that we reach through human reason is reliable. Feynman ends his with a ringing declaration of support for the encyclical *Pacem in Terris* issued by Pope John XXIII in 1963, in which the Pope called for peace among all nations based on truth, justice, charity, and liberty, and for the right organization of society to attain this end. Polkinghorne would certainly agree with Feynman's declaration. Feynman might not so readily agree with Polkinghorne's. Feynman believed that all human understanding is open to question. Even people who base their understanding on human reason sometimes make mistakes.

Postscript, 2006

In this review, when I wrote of "the religion of Jesus as we find it in his teachings recorded in the gospels," I had in mind the first three gospels. The fourth gospel, the Gospel of Saint John, shows us a very different Jesus, much more Greek in spirit, speaking about himself in the language of theology. So the clash between the Hebrew Jesus and the Greek Jesus already exists within the New Testament, between the first three gospels and the fourth. Saint John's gospel was certainly written later and was influenced by Greek ideas that probably came from Saint Paul. I am indebted to Elaine Pagels for conversations which taught me most of what I know about Christianity in general and Saint John's gospel in particular.

26

THIS SIDE IDOLATRY

"I DID LOVE the man this side idolatry as much as any," wrote the Elizabethan dramatist Ben Jonson. "The man" was Jonson's friend and mentor William Shakespeare. Jonson and Shakespeare were both successful playwrights. Jonson was learned and scholarly; Shakespeare was slapdash and a genius. There was no jealousy between them. Shakespeare was nine years older, already filling the London stage with masterpieces before Jonson began to write. Shakespeare was, as Jonson said, "honest and of an open and free nature," and gave his young friend practical help as well as encouragement. The most important help that Shakespeare gave was to act one of the leading roles in Jonson's first play, *Every Man in his Humour*, when it was performed in 1598. The play was a resounding success and launched Jonson's professional career. Jonson was then aged twenty-five, Shakespeare thirty-four. After 1598, Jonson continued to write poems and plays, and many of his plays were performed by Shakespeare's company. Jonson became famous in his own right as a poet and scholar, and at the end of his life he was honored with burial in Westminster Abbey. But he never forgot his debt to his old friend. When Shakespeare died, Jonson wrote a poem, "To the Memory of My Beloved Master, William Shakespeare," containing the well-known lines:

He was not of an age but for all time!...
Nature herself was proud of his designs
And joyed to wear the dressing of his lines....
Yet I must not give Nature all: thy art,
My gentle Shakespeare, must enjoy a part.
For though the poet's matter nature be,
His art does give the fashion; and, that he
Who casts to write a living line, must sweat,...
For a good poet's made, as well as born....

What have Jonson and Shakespeare to do with Richard Feynman? Simply this. I can say as Jonson said, "I did love this man this side idolatry as much as any." Fate gave me the tremendous luck to have Feynman as a mentor. I was the learned and scholarly student who came from England to Cornell University in 1947 and was immediately entranced by the slapdash genius of Feynman. With the arrogance of youth, I decided that I could play Jonson to Feynman's Shakespeare. I had not expected to meet Shakespeare on American soil, but I had no difficulty in recognizing him when I saw him.

Before I met Feynman, I had published a number of mathematical papers, full of clever tricks but totally lacking in importance. When I met Feynman, I knew at once that I had entered another world. He was not interested in publishing pretty papers. He was struggling, more intensely than I had ever seen anyone struggle, to understand the workings of nature by rebuilding physics from the bottom up. I was lucky to meet him near the end of his eight-year struggle. The new physics that he had imagined as a student of John Wheeler seven years earlier was finally coalescing into a coherent vision of nature, the vision that he called "the space-time approach." The vision was in 1947 still unfinished, full of loose ends and inconsistencies, but I saw at once that it had to be right. I seized every opportunity to listen to Feynman talk, to learn to swim in the deluge of his ideas. He loved to talk, and he welcomed me as a listener. So we became friends for life.

For a year I watched as Feynman perfected his way of describing nature with pictures and diagrams, until he had tied down the loose ends and removed the inconsistencies. Then he began to calculate numbers, using his diagrams as a guide. With astonishing speed he was able to calculate physical quantities that could be compared directly with experiment. The experiments agreed with his numbers. In the summer of 1948 we could see Jonson's words coming true: "Nature herself was proud of his designs/And joyed to wear the dressing of his lines."

During the same year when I was walking and talking with Feynman, I was also studying the work of the physicists Julian Schwinger and Sin-Itiro Tomonaga, who were following more conventional paths and arriving at similar results. Schwinger and Tomonaga had independently succeeded, using more laborious and complicated methods, in calculating the same quantities that Feynman could derive directly from his diagrams. Schwinger and Tomonaga did not rebuild physics. They took physics as they found it, and only introduced new mathematical methods to extract numbers from the physics. When it became clear that the results of their calculations agreed with Feynman, I knew that I had been given a unique opportunity to bring the three theories together. I wrote a paper with the title "The Radiation Theories of Tomonaga, Schwinger and Feynman," explaining why the theories looked different but were fundamentally the same. My paper was published in the *Physical Review* in 1949, and launched my professional career as decisively as *Every Man in his Humour* launched Jonson's. I was then, like Jonson, twenty-five years old. Feynman was thirty-one, three years younger than Shakespeare had been in 1598. I was careful to treat my three protagonists with equal dignity and respect, but I knew in my heart that Feynman was the greatest of the three and that the main purpose of my paper was to make his revolutionary ideas accessible to physicists around the world. Feynman actively encouraged me to publish his ideas, and never once complained that I was stealing his thunder. He was the chief actor in my play.

One of the treasured possessions that I brought from England to America was *The Essential Shakespeare* by J. Dover Wilson, a short biography of Shakespeare containing most of the quotations from Jonson that I have reproduced here.[1] Wilson's book is neither a work of fiction nor a work of history, but something in between. It is based on the firsthand testimony of Jonson and others, but Wilson used his imagination together with the scanty historical documents to bring Shakespeare to life. In particular, the earliest evidence that Shakespeare acted in Jonson's play comes from a document dated 1709, more than a hundred years after the event. We know that Shakespeare was famous as an actor as well as a writer, and I see no reason to doubt the traditional story as Wilson tells it.

Besides his transcendent passion for science, Feynman had also a robust appetite for jokes and ordinary human pleasures. Between his heroic struggles to understand the laws of nature, he loved to relax with friends, to play his bongo drums, to entertain everybody with tricks and stories. In this too he resembled Shakespeare. Out of Wilson's book I take the testimony of Jonson:

> When he hath set himself to writing, he would join night to day; press upon himself without release, not minding it till he fainted: and when he left off, remove himself into all sports and looseness again; that it was almost a despair to draw him to his book: but once got to it, he grew stronger and more earnest by the ease.

---

1. Cambridge University Press, 1932.

That was Shakespeare, and that was also the Feynman I knew and loved, this side idolatry.[2]

2. Since I wrote this foreword for the Feynman anthology, *The Pleasure of Finding Things Out: The Best Short Works of Richard Feynman*, edited by Jeffrey Robbins (Perseus, 1999), many more books about Feynman have been published. One of them is the volume of letters that I reviewed in Chapter 23. As he recedes into history, Feynman seems to rise higher and higher, like Shakespeare, above his contemporaries.

27

ONE IN A MILLION

DEBUNKED![1] IS SHORT and highly readable. It tells good stories about human foolishness masquerading as science. It offers useful assistance to citizens trying to tell the difference between sense and nonsense. When it was published in France,[2] the title was *Devenez sorciers, devenez savants*, which means literally, "Become magicians, become experts," or more freely translated, "Learn to do magic and learn to see through it." The English title misses the point. The book is saying that the best way to avoid being deceived by magic tricks is to learn to do the tricks yourself.

The translator is a medical school faculty member who has written books about probability theory for doctors designing clinical trials. His knowledge of French language and culture was acquired from his family. He has added a preface explaining how he dealt with the problems of translation. He says, "Very long sentences in a distinctive, glorious Gallic rhetorical style have been reduced in frequency by rewriting but not eliminated completely.... I opted for intelligibility rather than the unalloyed preservation of style." Few of his sentences

1. Georges Charpak and Henri Broch, *Debunked! ESP, Telekinesis, and Other Pseudoscience*, translated from the French by Bart K. Holland (Johns Hopkins University Press, 2004).

2. Paris: Éditions Odile Jacob, 2002.

are painfully long, and none is unintelligible. What remains of the distinctive Gallic rhetorical style is to be found in some passages where the authors express a lofty contempt for those who disagree with them. "Is it acceptable," they ask, "that university colleagues, due to laziness, lack of rigor, lack of competence, or love of media attention, should go along with a pack of errors, untruths, nonsense, or lies, and label it an honorable point of view?" The translator has retained enough of the elegant style to give us the flavor of the French original. But the polemical passages are few. The greater part of the book is straight storytelling. The stories are well told and are allowed to speak for themselves.

The name of Georges Charpak brings back memories of fifty years ago, when I was living on the side of a mountain above the village of Les Houches, in the high alpine region of France close to Mont Blanc. I was teaching physics at the Les Houches Summer School, an institution that was then three years old and is now still flourishing. It was founded by Cécile DeWitt, who was then a young postdoctoral student, with the avowed purpose of rejuvenating French physics. Cécile is no longer running the school, but she is still very much alive and helping to keep the enterprise going. In 1951, when she founded the school, theoretical physics in France was at a low ebb, with academic jobs in the leading universities monopolized by old men out of touch with new developments. Cécile raised her own meager funds and built her school in a faraway corner of France, out of reach of the mandarins in Paris. She bought an abandoned farm and made the buildings more or less habitable as best she could. Students flocked to the school from all over Europe. The class of students that I taught that summer were the best I ever had. Many of them later became famous as scientific leaders in their own countries. The brightest of all was Georges Charpak.

We lived together in a cowshed and the students listened to lectures in a barn. I gave a tough course with the title Advanced Quantum

Mechanics. I worked hard teaching and the students worked hard learning. But the formal lectures were the least important part of the school. Much more memorable were the informal sessions, the meals and the hikes, the daily hardships of mud and rain that we all shared. In a few short years the school became a prime mover of the renaissance of physics, not only in France but all over Europe. It has continued to be a center of excellence, bringing together gifted young people and giving them the opportunity to work together and learn together, creating friendships that last a lifetime. The cowsheds have long ago been replaced by solid permanent buildings, the muddy farmyard by a terrace ornamented with modern sculpture. In the year 1954 when I was at Les Houches, all of us, Cécile and the lecturers and the students, were young. We were intoxicated with joie de vivre. We were Europeans, we had survived the war and the dismal years of impoverishment that followed it, and now we finally saw Europe rising from the ashes and rebuilding itself. Les Houches was a visible symbol of the rebuilding, which was spiritual as well as physical. We knew we were lucky to be a part of it.

Somewhere on the farm, Georges found an old skull of a bull with horns attached, and he liked to wear these horns on his head. The horns fitted well with his bull-like physique and character. I have a vivid memory of Georges roaring around a muddy field with the horns, pretending to be a bull. His newlywed wife, Dominique, armed with a long spoon from the school kitchen, was pretending to be a matador. For most of his life, Georges has been a leader of experiments at CERN, the European Council for Nuclear Research, on the border between France and Switzerland. CERN is not far from Les Houches and is another symbol of the scientific rejuvenation of Europe. It came as no surprise in 1992 when I heard that Georges was the first of our students to win a Nobel Prize in Physics. It comes as no surprise to meet him again in this book, fighting fiercely against the enemies of scientific reason.

Henri Broch, the second author of this book, is a professor of physics at the University of Nice. He is not as famous as Charpak as a physicist, but he is famous as an investigator of paranormal phenomena such as extrasensory perception and telepathy. He has investigated many claims of people who believe that they possess paranormal powers. His success in demolishing paranormal claims is owing to his skill as a magician. He has mastered the art of doing magical tricks, so that he can reproduce the allegedly paranormal phenomena in public demonstrations.

Broch plays the same role in France that the Amazing Randi plays in America. Like Broch, the Amazing Randi is a skilled magician. He challenges any possessor of psychic powers to perform wonders that he cannot duplicate. I once participated in a session in San Diego at which the Amazing Randi confronted the famous Israeli spoon-bender Uri Geller. Geller gave a public performance at which he bent metal objects such as spoons and keys without touching them, using his psychic powers. To make his performances more impressive, he used the word "telekinesis" to describe what ordinary people call spoon-bending. The session was conducted in a big public auditorium, and a large crowd came. I took my family along to see the show.

Following his usual routine, Geller invited volunteers from the audience to come onstage, bringing spoons or keys with them. My daughter Emily, then twelve years old, volunteered and went up with an old key that we had brought with us. The key was no longer in use, and we did not mind if Geller succeeded in bending it. Geller examined the key carefully and handed it back to Emily, telling her to hold on to it and not let go. Then he chatted with the audience about telekinesis. He described to us how the atoms in the key were rearranging themselves in response to his psychic powers, while Emily stood waiting on the stage. Then he turned suddenly to Emily and said, "Now, let's look at your key." She handed him the key, and there it was, bent. He gave it back to her, and she came down to show it to the audience. She said she could have sworn that she was watching the key the whole time and never let it out of her sight. After that, Geller continued with other volunteers, bending various other objects. Then Geller departed and Randi's performance began.

Randi went through the same rituals as Geller and was equally successful. A succession of volunteers went onstage, and came down as mystified as Emily had been. Then Randi explained how he did it. The actual bending of the key was the easy part. He bent it with one hand, inserting the tip into a hole in a second key and squeezing the two keys together. This was done while he was chatting with the audience about psychic powers. The more difficult part of the trick was the exchange of keys. The exchange had to be done twice. At the beginning when the volunteer first handed him the key, and at the end when he gave it back to her, he quickly exchanged her key for a similar key that he had hidden in his hand. Each time he made the exchange, he distracted her attention and the attention of the audience with a loud remark or a sudden movement of the other hand. The essential skill of the magician is the ability to distract attention at the moment when the trickery is done. Emily said that if she had not seen Randi's demonstration she would not have believed that she could be so easily deceived. After Randi had reproduced all of Geller's tricks and explained how they were done, he entertained the audience with a number of even more amazing tricks which he did not explain.

Broch describes, in a section of the book entitled "Practice Telepathy," a splendid demonstration of his ability to send information telepathically to an accomplice a few miles away. The demonstration is made in a private house where a group of friends, in no way involved in the trickery, are gathered. First Broch invites the group to provide a deck of cards, to shuffle them thoroughly, and to pick a card at random. Broch sits in the room while this is done, so it is obvious that he could have had no influence on the picking of the card, and no prior knowledge of which card would be picked. Let us suppose that the five of clubs is picked. Broch then glances at an address book that he carries in his pocket, writes down the name and telephone number of the accomplice on a piece of paper, and asks the friends to choose a representative who goes to another room to call the number. While the call is made, Broch sits in his chair gazing at the card with a look of intense concentration, groaning with the effort of exercising his telepathic powers. The accomplice answers the phone and says, "We have several brothers living here, which of us do you want to talk to?" So the friend gives the name and the accomplice says, "Speaking." The friend explains that a card has been picked and that Broch is trying to transmit the image of the card telepathically. After a suitable pause for dramatic effect, the accomplice says, "The card you picked is the five of clubs," and the friend rushes back to tell Broch and the rest of the assembled company that the message has got through.

Hardly anyone who witnesses this performance and is not an expert magician can see through it. To the uninitiated it looks like good solid scientific evidence for telepathy. The essential clue, which almost everyone misses, is the address book that Broch glances at before writing down the name of the accomplice. The address book contains a list of the fifty-three cards in a standard deck, each paired with a common French first name. The accomplice has another copy of the same list. As soon as the accomplice hears the name, he knows the card.

The book also has a good chapter on "Amazing Coincidences." These are strange events which appear to give evidence of supernatural influences operating in everyday life. They are not the result of deliberate fraud or trickery, but only of the laws of probability. The paradoxical feature of the laws of probability is that they make unlikely events happen unexpectedly often. A simple way to state the paradox is Littlewood's law of miracles. John Littlewood was a famous mathematician who was teaching at Cambridge University when I was a student. Being a professional mathematician, he defined miracles precisely before stating his law about them. He defined a miracle as an event that has special significance when it occurs, but occurs with a probability of one in a million. This definition agrees with our commonsense understanding of the word "miracle."

Littlewood's law of miracles states that in the course of any normal person's life, miracles happen at a rate of roughly one per month. The proof of the law is simple. During the time that we are awake and actively engaged in living our lives, roughly for eight hours each day, we see and hear things happening at a rate of about one per second. So the total number of events that happen to us is about 30,000 per day, or about a million per month. With few exceptions, these events are not miracles because they are insignificant. The chance of a miracle is about one per million events. Therefore we should expect about one miracle to happen, on the average, every month. Broch tells stories of some amazing coincidences that happened to him and his friends, all of them easily explained as consequences of Littlewood's law.

A large number of people calling themselves parapsychologists have tried to study paranormal phenomena using rigorous scientific methods. Their favorite tool is a little deck of twenty-five cards, with one of five symbols on each card. The five symbols are squares, circles, stars, crosses, and squiggles. An ideal telepathy experiment is done with two people, the sender and the receiver, sitting in separate rooms, with careful controls to eliminate all possibility of communication between the two of them or between them and the experimenter. The sender and the receiver synchronize their activities with accurate clocks. At fixed times agreed in advance, the sender picks cards from a well-shuffled deck, gazes at them one at a time, and records the sequence of cards gazed at. At the same times, the receiver guesses the cards and records the sequence of guesses. At the end of the experiment, an impartial witness, not the experimenter, compares the two records and finds the percentage of correct guesses. If telepathy is not operating, the percentage should be close to twenty. If the percentage is persistently higher than twenty, the experimenter may claim to have found evidence for telepathy.

If this idealized picture of a telepathy experiment were real, we should long ago have been able to decide whether telepathy exists or not. In the real world, the way such experiments are done is very different, as I know from personal experience. When I was a teenager long ago, parapsychology was fashionable. I bought a deck of parapsychology cards and did card-guessing experiments with my friends. We spent long hours, taking turns at gazing and guessing cards. Unlike Broch, we were strongly motivated to find positive evidence of telepathy. We considered it likely that telepathy existed and we wanted to prove ourselves to be telepathically gifted. When we started our sessions, we achieved some spectacularly high percentages of correct guesses. Then, as time went on, the percentages declined toward twenty and our enthusiasm dwindled. After a few months of sporadic efforts, we put the cards away and forgot about them.

Looking back on our experience with the cards, we came to understand that there are three formidable obstacles to any scientific study of telepathy. The first obstacle is boredom. The experiments are insufferably boring. In the end we gave up because we could not stand the boredom of sitting and guessing cards for hours on end. The second obstacle is inadequate controls. We never even tried to impose rigorous controls on communication between sender and receiver. Without such controls, our results were scientifically worthless. But any serious system of controls, stopping us from chatting and joking while we were gazing and guessing, would have made the experiments even more insufferably boring.

The third obstacle is biased sampling. The results of such experiments depend crucially on when you decide to stop. If you decide to stop after the initial spectacularly high percentages, the results are strongly positive. If you decide to stop when you are almost dying of boredom, the results are strongly negative. The only way to obtain unbiased results is to decide in advance when to stop, and this we had not done. We were not disciplined enough to make a decision in advance to do 10,000 guesses and then stop, regardless of the percentage of correct guesses that we might have achieved. We did not succeed in overcoming a single one of the three obstacles. To reach any scientifically credible conclusions, we would have needed to overcome all three.

The history of the card-guessing experiments, carried out initially by Joseph Rhine at Duke University and later by many other groups following Rhine's methods, is a sorry story. A number of experiments that claimed positive results were later proved to be fraudulent. Those that were not fraudulent were plagued by the same three obstacles that frustrated our efforts. It is difficult, expensive, and tedious to impose controls rigorous enough to eliminate the possibility of fraud. And even after such controls have been imposed, the conclusions of a series of experiments can be strongly biased by selective reporting of the results. Littlewood's law applies to experimental results as well as to the events of daily life. A session with a noticeably high percentage of correct guesses is a miracle according to Littlewood's definition. If a large number of experiments are done by various groups under various conditions, miracles will occasionally occur. If miracles are selectively reported, they are experimentally indistinguishable from real occurrences of telepathy.

Charpak and Broch see the modern growth of astrology and other pseudosciences as a rising menace that they are called upon to fight to the death. They are horrified by the prevalence of unscientific thinking among students in France today. They sum up their response to these irrationalities in another elegant Gallic sentence: "The question is inescapable: Isn't scientific thought the indispensable companion to wisdom, to clear thinking, and to the love of those virtues, which is expressed not only in vain incantations to the sky but also in logical actions?" They have here touched on a question which goes to the heart of the matter. What are the proper limits of science?

There are two extreme points of view concerning the role of science in human understanding. At one extreme is the reductionist view, holding that all kinds of knowledge, from physics and chemistry to psychology and philosophy and sociology and history and ethics and religion, can be reduced to science. Whatever cannot be reduced to science is not knowledge. The reductionist view was forcibly expressed by Edward Wilson in his 1998 book *Consilience*. At the other extreme is the traditional view, that knowledge comes from many independent sources, and science is only one of them. Knowledge of good and evil, knowledge of grace and beauty, knowledge of ethical and artistic values, knowledge of human nature derived from history and literature or from intimate acquaintance with family and friends, knowledge of the nature of things derived from meditation or from religion, all are sources of knowledge that stand side by side with science, parts of a human heritage that is older than science and perhaps more enduring. Most people hold views intermediate between the two extremes. Charpak and Broch are close to the reductionist extreme, while I am close to the traditional extreme.

The question of the proper limits of science has a strong connection with the possible existence of paranormal phenomena. Charpak and Broch and I agree that attempts to study extrasensory perception and telepathy using the methods of science have failed. Charpak and Broch say that since extrasensory perception and telepathy cannot be studied scientifically, they do not exist. Their conclusion is clear and logical, but I do not accept it because I am not a reductionist. I claim that paranormal phenomena may really exist but may not be accessible to scientific investigation. This is a hypothesis. I am not saying that it is true, only that it is tenable, and to my mind plausible.

The hypothesis that paranormal phenomena are real but lie outside the limits of science is supported by a great mass of evidence.

The evidence has been collected by the Society for Psychical Research in Britain and by similar organizations in other countries. The journal of the London society is full of stories of remarkable events in which ordinary people appear to possess paranormal abilities. The evidence is entirely anecdotal. It has nothing to do with science, since it cannot be reproduced under controlled conditions. But the evidence is there. The members of the society took great trouble to interview firsthand witnesses as soon as possible after the events, and to document the stories carefully. One fact that emerges clearly from the stories is that paranormal events occur, if they occur at all, only when people are under stress and experiencing strong emotion. This fact would immediately explain why paranormal phenomena are not observable under the conditions of a well-controlled scientific experiment. Strong emotion and stress are inherently incompatible with controlled scientific procedures. In a typical card-guessing experiment, the participants may begin the session in a high state of excitement and record a few high scores, but as the hours pass, and boredom replaces excitement, the scores decline to the 20 percent expected from random chance.

I am suggesting that paranormal mental abilities and scientific method may be complementary. The word "complementary" is a technical term introduced into physics by Niels Bohr. It means that two descriptions of nature may both be valid but cannot be observed simultaneously. The classic example of complementarity is the dual nature of light. In one experiment light is seen to behave as a continuous wave, in another experiment it behaves as a swarm of particles, but we cannot see the wave and the particles in the same experiment. Complementarity in physics is an established fact. The extension of the idea of complementarity to mental phenomena is pure speculation. But I find it plausible that a world of mental phenomena should exist, too fluid and evanescent to be grasped with the cumbersome tools of science.

I should here declare my personal interest in the matter. One of my grandmothers was a notorious and successful faith healer. One of my cousins was for many years the editor of the *Journal of the Society for Psychical Research*. Both these ladies were well educated, highly intelligent, and fervent believers in paranormal phenomena. They may have been deluded, but neither of them was a fool. Their beliefs were based on personal experience and careful scrutiny of evidence. Nothing that they believed was incompatible with science.

Whether paranormal phenomena exist or not, the evidence for their existence is corrupted by a vast amount of nonsense and outright fraud. Before we can begin to evaluate the evidence, we must get rid of the hucksters and charlatans who have turned unsolved mysteries into a profitable business. Charpak and Broch have done a fine job, sweeping out the money-changers from the temple of science and exposing their tricks. I recommend this book to believers and skeptics alike. It is good entertainment, whether or not you believe in astrology.

Postscript, 2006

A deluge of eloquent letters came in response to this review. Orthodox scientists were outraged because I considered the existence of telepathy to be possible. True believers in telepathy were outraged because I considered its existence to be unproved. This is a question that is of deep concern to many readers. The most interesting response came from Rupert Sheldrake, who sent me papers describing his experiments studying telepathy in dogs. Dogs have several advantages over humans as experimental subjects. They do not get bored, they do not cheat, and they do not have any interest in the outcome of the experiment. Sheldrake's experiments contradict my statement that telepathy cannot be studied scientifically. Unfortunately, the experiments were conducted by humans, not by dogs, and effects of human bias and selective reporting could not be altogether eliminated. But Sheldrake is right when he says that the experiments are scientific. They are repeatable, and ought to be repeated by independent investigators using different dogs. Interested readers may examine the evidence in Sheldrake's book, *Dogs That Know When Their Owners Are Coming Home, and Other Unexplained Powers of Animals* (Crown, 1999), and in his more recently published papers.

28

MANY WORLDS

LIFE SOMETIMES IMITATES art. Olaf Stapledon wrote *Star Maker* sixty-six years ago as a dramatization of a philosophical idea. Now, sixty-six years later, cosmologists are proposing similar scenarios as possible models of the universe we live in.

Stapledon was a philosopher and not a scientist. He wrote this book to explore an elegant new solution of the old philosophical problem of evil. The problem is to reconcile the existence of evil in our world with the existence of an omnipotent and not entirely malevolent creator. The solution is to suppose that our universe is only one of many, that the creator is engaged in creating a long series of universes, that he is improving his designs as he goes along, that our universe is one of his early flawed creations, and that the evils that we see around us are flaws from which the creator will learn how to do the job better next time. Stapledon brings the story to a climax in his penultimate chapter, "The Maker and His Works," which paints a powerful picture of the creator as a craftsman using us as raw material to practice his skills. The hero of the story is a human observer, who first explores the multitude of worlds in our universe on which intelligent life has evolved, and then finally confronts the Star Maker. But that supreme moment of confrontation is tragic rather than harmonious. Like God answering Job out of the whirlwind, the Star

Maker strikes him down and rejects him. The Star Maker judges his creation with love but without mercy. In the end, our entire universe, in spite of all its majesty and beauty, is a flawed experiment. The Star Maker is already busy with designs for other universes in which our flaws may be repaired.

Within the last few years, Lee Smolin and Martin Rees and other cosmologists have been addressing a very different philosophical problem. They are not concerned with moral judgments but with scientific facts. It is a scientific fact that our universe is strangely hospitable to the growth of life and intelligence. Many details of the laws of physics and chemistry seem to conspire to make our universe friendly to life. If the details of nuclear forces, gravitational forces, and chemical forces had been slightly different, life as we know it could never have evolved. Our form of life was able to adapt itself to the universe in which it originated, but there is no way in which any form of life could have adapted itself to a universe that collapsed into a fiery space-time singularity before it had time to give birth to stars and planets, or to a universe that expanded into a cold dilute gas before giving birth to any atoms heavier than hydrogen. It seems that our universe is somehow fine-tuned, with the forces of gravitational collapse and cosmic expansion almost exactly balanced, so that stars and planets have time to evolve, and the atoms of the various chemical elements necessary for life have time to be formed. The philosophical problem of explaining why the universe is hospitable to life is called the fine-tuning problem.

Lee Smolin was the first cosmologist to point out that the fine-tuning problem is solved if we assume that our universe is one of many, that new universes sprout like babies within older universes, that babies resemble their parents, that baby universes are born with a random assortment of laws of physics and chemistry, and that longer-lived universes give birth to more numerous babies. When these assumptions are made, it follows that a process of Darwinian evolution will select longer-lived universes. And it is no accident that creatures like us happen to live in one of the longer-lived universes that has laws of physics and chemistry fine-tuned to the growth of life and intelligence. Among the billions of universes that exist, only a few will be fine-tuned enough for life to evolve, and one of those few must be our home. Smolin's many-universe cosmology does not require a Star Maker to design it. It only needs the hidden hand of natural selection to guide its evolution, and the luck of a favorable random variation to provide a home for life.

The astronomer Martin Rees has suggested another way in which nature might have solved the fine-tuning problem. He remarks that if a multitude of universes exists, then some of them are likely to allow the evolution of life-forms with mental processes far more advanced than ours. A superintelligent life-form might be capable of simulating in its brain or in a supercomputer the complete history of another universe with a lower degree of complexity. Now Rees asks the question: Might we and the universe we live in be simulations, lacking any real physical substance and only existing as mental constructions in the minds of our superintelligent colleagues? "This concept," says Rees, "opens up the possibility of a new kind of virtual time travel, because the advanced beings creating the simulation can, in effect, rerun the past. It's not a time-loop in a traditional sense: it's a reconstruction of the past, allowing advanced beings to explore their history." Rees solves the fine-tuning problem by allowing future superintelligent beings to fine-tune the past.

If our present universe is a simulation created by intelligent aliens interested in exploring the consequences of alternative laws of physics, then we should expect the laws of physics that we observe to be chosen in such a way as to make our universe as interesting as possible. We should expect to find our universe allowing structures and processes of maximum diversity. The immense richness of ecological environments on our own planet gives support to Rees's proposal. So does the diversity of planets, stars, galaxies, and other more exotic inhabitants of the celestial zoo. We observe our universe to be not only friendly to life, but friendly to the maximum diversity of life. As the biologist J. B. S. Haldane once remarked, the Creator appears to have an inordinate fondness for beetles. Future beings with superior intelligence may be fond of beetles too.[1]

Now comes the big surprise. The proposal of Rees, imagining our universe to be a simulation in the minds of intelligent aliens, is very similar to the proposal of Stapledon, imagining our universe to be an artifact of the Star Maker. The two proposals were addressed to different philosophical problems, the problem of evil for Stapledon and the problem of fine-tuning for Rees. They arose out of different cultures, the culture of moral philosophy for Stapledon and the culture of scientific cosmology for Rees. Yet, coming from different cultures and different concerns, Stapledon and Rees converged upon the same solution. This is not the first time that a writer of science fiction has leapt ahead to a new vision of reality that only became a part of the discourse of respectable scientists half a century later. Stapledon's Star Maker is like Jules Verne's Captain Nemo and H. G. Wells's Doctor Moreau, fictional characters who have become more real as the years go by.

Another fictional character who has become more real as the years go by is Sirius, the superintelligent dog who is the hero of Stapledon's next major novel after *Star Maker*. Stapledon published *Sirius* in 1944. It is another story presenting a philosophical problem in dramatic form. The philosophical problem addressed by *Sirius* is the moral status of animals. Since animals evidently have feelings and

---

1. I am not sure whether the statement that the Creator has an inordinate fondness for beetles originated with Haldane or with Darwin. Pat McCarthy, the editor of this new edition of *Star Maker*, informs me that *The Oxford Dictionary of Quotations* attributes the statement to Haldane. It is still possible that Haldane cribbed it from Darwin.

emotions similar to ours, why do they not have legal rights and moral standing similar to ours? In the half-century since *Sirius* was written, campaigns for animal rights and animal liberation have gained strength in many countries. As we have learned more about the rich social and emotional lives of animals living in the wild, our denial of rights to animals living in captivity becomes harder to justify. And the rapid progress of genetic engineering will make the emergence of superintelligent dogs and cats a real possibility in the century now beginning. Stapledon's story gives us warning of the tragic consequences that may follow from a blurring of the boundaries of species.

I hope that Pat McCarthy will edit a new edition of *Sirius* as a companion to his edition of *Star Maker*. In my opinion, *Sirius* is the finest of all Stapledon's works. It received less attention than it deserved when it was published, because the majority of readers in 1944 were preoccupied with more urgent matters. In style it is less didactic and more personal than *Star Maker*. The drama of *Sirius* is played out against a vivid background, the sheepdog country of North Wales which Stapledon knew and loved. The characters are real human beings and dogs rather than disembodied spirits. The tragedy is the predicament of a lonely creature who understands both the world of dogs and the world of humans but belongs to neither.

But I must not praise *Sirius* too highly. That would be unfair to *Star Maker*. *Star Maker* may be, like the universe we happen to live in, a flawed masterpiece, but it is still a masterpiece. It is a classic work of imaginative literature, speaking to our modern age. It should be on the list of Great Books that anyone claiming to be educated should read. It is worthy to be compared, as McCarthy compares it in the introduction following this preface, with *The Divine Comedy* of Dante.

29

RELIGION FROM THE OUTSIDE

BREAKING THE SPELL of religion is a game that many people can play. The best player of this game that I ever knew was Professor G.H. Hardy, a world-famous mathematician who happened to be a passionate atheist. There are two kinds of atheists, ordinary atheists who do not believe in God and passionate atheists who consider God to be their personal enemy. When I was a junior fellow at Trinity College, Cambridge, Hardy was my mentor. As a junior fellow I enjoyed the privilege of dining at the high table with the old and famous. During my tenure, Professor Simpson, one of the old and famous fellows, died. Simpson had a strong sentimental attachment to the college and was a religious believer. He left instructions that he should be cremated and his ashes should be scattered on the bowling green in the fellows' garden where he loved to walk and meditate. A few days after he died, a solemn funeral service was held for him in the college chapel. His many years of faithful service to the college and his exemplary role as a Christian scholar and teacher were duly celebrated.

In the evening of the same day I took my place at the high table. One of the neighboring places at the table was empty. Professor Hardy, contrary to his usual habit, was late for dinner. After we had all sat down and the Latin grace had been said, Hardy strolled into the dining hall, ostentatiously scraping his shoes on the wooden floor and complaining in a loud voice for everyone to hear, "What is this awful stuff they have put on the grass in the fellows' garden? I can't get it off my shoes." Hardy, of course, knew very well what the stuff was. He had always disliked religion in general and Simpson's piety in particular, and he was taking his opportunity for a little revenge.

Paul Erdös was another world-famous mathematician who was a passionate atheist. Erdös always referred to God as SF, short for Supreme Fascist. Erdös had for many years successfully outwitted the dictators of Italy, Germany, and Hungary, moving from country to country to escape from their clutches. He called his God SF because he imagined God to be a fascist dictator like Mussolini, powerful and brutal but rather slow-witted. Erdös was able to outwit SF by moving frequently from one place to another and never allowing his activities to fall into a predictable pattern. SF, like the other dictators, was too stupid to understand Erdös's mathematics. Hardy and Erdös were both lovable characters, contributing more than their fair share to the human comedy. Both of them were gifted clowns as well as great mathematicians.

And now comes Daniel Dennett to take his turn at breaking the spell. Dennett is a philosopher. In *Breaking the Spell: Religion as a Natural Phenomenon*[1] he is confronting the philosophical questions arising from religion in the modern world. Why does religion exist? Why does it have such a powerful grip on people in many different cultures? Are the practical effects of religion preponderantly good or preponderantly evil? Is religion useful as a basis for public morality? What can we do to counter the spread of religious movements that we consider dangerous? Can the tools and methods of science help us to understand religion as a natural phenomenon? Dennett remarks at the beginning that he will proceed

1. Viking Penguin, 2006.

> not by answering the big questions that motivate the whole enterprise but by asking them, as carefully as I can, and pointing out what we already know about how to answer them, and showing why we need to answer them.
>
> I am a philosopher, not a biologist or an anthropologist or a sociologist or historian or theologian. We philosophers are better at asking questions than at answering them....

Dennett practices what he preaches. He does not answer the questions, but takes four hundred pages to ask them. The book proceeds at a leisurely pace, with an easy conversational style and many digressions. It is divided into three sections, the first concerned with the nature of scientific inquiry, the second concerned with the history and evolution of religion, the third concerned with religion as it exists today. In the first section, Dennett defines scientific inquiry in a narrow way, restricting it to the collection of evidence that is reproducible and testable. He makes a sharp distinction between science on the one hand and the humanistic disciplines of history and theology on the other. He does not accept as scientific the great mass of evidence contained in historical narratives and personal experiences. Since it cannot be reproduced under controlled conditions, it does not belong to science. He quotes with approval and high praise several passages from *The Varieties of Religious Experience*, the classic description of religion from the point of view of a psychologist, published by William James in 1902. He describes James's book as "a treasure trove of insights and arguments, too often overlooked in recent times." But he does not accept James's insights and arguments as scientific.

James is examining religion from the inside, like a doctor trying to see the world through the eyes of his patients. James was trained as a medical doctor before he became a professor of psychology. He studied the personal experiences of saints and mystics as evidence of something real existing in a spiritual world beyond the boundaries of space and time. Dennett honors James as an explorer of the human condition, but not as an explorer of a spiritual world. For Dennett, the visions of saints and mystics are worthless as evidence, since they are neither repeatable nor testable. Dennett is examining religion from the outside, following the rules of science. For him, the visions of saints and mystics are only a phenomenon to be explained, like falling in love or hating people of a different skin color, mental conditions that may or may not be considered pathological.

The second section of the book is the longest and contains the core of Dennett's argument. He describes the various stages of the long historical evolution of religion, beginning with primitive tribal myths and rituals, and ending with the market-driven evangelical megachurches of modern America. Looking at these evolutionary processes from the outside, he speculates about ways in which they might be understood scientifically. He explains them tentatively as products of a Darwinian competition between belief systems, in which only the fittest belief systems survive. The fitness of a belief system is defined by its ability to make new converts and retain their loyalty. It has little to do with the biological fitness of its human carriers, and it has nothing to do with the truth or falsehood of the beliefs. Dennett emphasizes the fact that his explanation of the evolution of religion is testable with the methods of science. It could be tested by quantitative measurements of the transmissibility and durability of various belief systems. These measurements would provide an objective scientific test, to find out whether the surviving religions are really fitter than those that became extinct.

Dennett puts forward other hypotheses concerning the evolution of religion. He observes that belief, which means accepting certain doctrines as true, is different from belief in belief, which means believing belief in the same doctrines to be desirable. He finds evidence that large numbers of people who identify themselves as religious believers do not in fact believe the doctrines of their religions but only believe in belief as a desirable goal. The phenomenon of "belief in belief" makes religion attractive to many people who would otherwise be hard to convert. To belong to a religion, you do not have to believe. You only have to want to believe, or perhaps you only have to pretend to believe. Belief is difficult, but belief in belief is easy. Belief in belief is one of the important phenomena that give a religion increased transmissibility and consequently increased fitness. Dennett puts forward this connection between belief in belief and fitness as a hypothesis to be tested, not as a scientifically established fact. He regrets that little of the relevant research has yet been done. The title *Breaking the Spell* expresses his hope that when the scientific analysis of religion has been completed, the power of religion to overawe human reason will be broken.

Dennett has an easy time poking fun at the modern evangelical megachurches which pay more attention to the size of their congregations than to the quality of their religious life. The leaders of these churches are selling their versions of religion in a competitive market, and those that have the best marketing skills prevail. The market favors practical convenience rather than serious commitment to a pure and holy life. Looking at religion from the outside, Dennett sees clearly how the leaders of religious organizations are corrupted by power and money. He quotes Alan Wolfe, one of the sociologists who study American religious organizations and practices:

> Evangelicalism's popularity is due as much to its populistic and democratic urges—its determination to find out exactly what believers want and to offer it to them—as it is to certainties of the faith.... The term "sanctuary" is shunned by one church because of its "strong religious connotations," and more attention is paid to providing plenty of free parking and babysitting than to the proper interpretation of passages of Scripture.

Like Hardy and Erdös, Dennett plays the game of breaking the spell by making religion look silly. Many of my scientist friends and colleagues have similar prejudices. One famous scientist for whom I have a deep respect said to me, "Religion is a childhood disease from which we have recovered." There is nothing wrong with such prejudices, provided that they are openly admitted. Dennett's account of the evolution of religion is on the whole fair and well balanced.

The third and last section of Dennett's book describes his view of religion in the modern world. In a long chapter entitled "Morality and Religion," he blames religion for many of the worst evils of our century. He blames not only the minority of murderous fanatics whose religion impels them to acts of terrorism but also the majority of peaceful and moderate believers who do not publicly condemn the actions of the fanatics. This is a serious problem, whether one is dealing with Irish Catholic fanatics in Belfast or with Muslim fanatics in Britain and Spain. He quotes with approval the famous remark of the physicist Stephen Weinberg: "Good people will do good things, and bad people will do bad things. But for good people to do bad things—that takes religion." Weinberg's statement is true as far as it goes, but it is not the whole truth. To make it the whole truth, we must add an additional clause: "And for bad people to do good things—that takes religion." The main point of Christianity is that it is a religion for sinners. Jesus made that very clear. When the Pharisees asked his disciples, "Why eateth your Master with publicans and sinners?" he said, "I come to call not the righteous but sinners to repentance." Only a small fraction of sinners repent and do good things, but only a small fraction of good people are led by their religion to do bad things.

I see no way to draw up a balance sheet, to weigh the good done by religion against the evil and decide which is greater by some impartial process. My own prejudice, looking at religion from the inside, leads me to conclude that the good vastly outweighs the evil. In many places in the United States, with widening gaps between rich and poor, churches and synagogues are almost the only institutions that bind people together into communities. In church or in synagogue, people from different walks of life work together in youth groups or adult education groups, making music or teaching children, collecting money for charitable causes, and taking care of one another when sickness or disaster strikes. Without religion, the life of the country would be greatly impoverished. I know nothing at first hand about Islam, but by all accounts the mosques in Islamic countries, and to some extent in America too, play a similar role in holding communities together and taking care of widows and orphans.

Dennett, looking at religion from the outside, comes to the opposite conclusion. He sees the extreme religious sects that are breeding grounds for gangs of young terrorists and murderers, with the mass of ordinary believers giving them moral support by failing to turn them in to the police. He sees religion as an attractive nuisance in the legal sense, meaning a structure that attracts children and young people and exposes them to dangerous ideas and criminal temptations, like an unfenced swimming pool or an unlocked gun room. My view of religion and Dennett's are equally true and equally prejudiced. I see religion as a precious and ancient part of our human heritage. Dennett sees it as a load of superfluous mental baggage which we should be glad to discard.

After Dennett's harsh depiction of the moral evils associated with religion, his last chapter, "Now What Do We Do?," is bland and conciliatory. "So, in the end," he says, "my central policy recommendation is that we gently, firmly educate the people of the world, so that they can make truly informed choices about their lives." This recommendation sounds harmless enough. Why can we not all agree with it? Unfortunately, it conceals fundamental disagreements. To give the recommendation a concrete meaning, the meaning of the little word "we" must be specified. Who are the "we" who are to educate the people of the world? At stake is the political control of religious education, the most contentious of all the issues that religion poses to modern societies. "We" might be the parents of the children to be educated, or a local school board, or a national ministry of education, or a legally established ecclesiastical authority, or an international group of philosophers sharing Dennett's views. Of all these possibilities, the last is the least likely to be implemented. Dennett's recommendation leaves the practical problems of regulating religious education unsolved. Until we can agree about the meaning of "we," the recommendation to "gently, firmly educate the people of the world" will only cause further dissension between religious believers and well-meaning philosophers.

The control of education is the arena in which political fights between religious believers and civil authorities become most bitter. In the United States these fights are made peculiarly intractable by the legal doctrine of separation of church and state, which forbids public schools to provide religious instruction. Parents with fundamentalist beliefs have a legitimate grievance, being compelled to pay for public schools which they see as destroying the religious faith of their children. This feeling of grievance was avoided in England through the wisdom of Thomas Huxley, a close friend of Charles Darwin and a leading proponent of Darwin's theory of evolution. When public education was instituted in England in 1870, eleven years after Darwin's theory was published, Huxley was appointed to the royal commission which decided what to teach in the public schools.

Huxley was himself an agnostic, but as a member of the commission he firmly insisted that religion should be taught in schools together with science. Every child should be taught the Christian Bible as an integral part of English culture. In recent times the scope of religious instruction in England has been extended to include Judaism and Islam. As a result of this policy, no strong antagonism between religious parents and public schools has arisen, from 1870 until the present day. The teaching of religion in public schools coincided with a decline of religious belief and a growth of religious tolerance. Children exposed to religion in public schools do not as a rule take it seriously. We do not know whether Huxley foresaw the decline of religion in England, but there is no doubt that he would have welcomed this unintended consequence of his educational policy.

It is unfortunate that Huxley's solution of the problem of religious education is not available to the United States. Every country is different, especially in matters concerning religion, and no single solution to the problem of religious education fits all. In each country, a workable solution has to be found by political compromise between conflicting views, within the rules imposed by the local culture. To be workable, a solution does not need to be scientifically or philosophically consistent. When I was a boy in England long ago, people who traveled on trains with dogs had to pay for a dog ticket. The question arose whether I needed to buy a dog ticket when I was traveling with a tortoise. The conductor on the train gave me the answer: "Cats is dogs and rabbits is dogs but tortoises is insects and travel free according." The rules governing religious education should be administered with a similar freedom of interpretation.

Dennett also advocates more intensive research on religion considered from a scientific point of view. Here again, we can all agree with the recommendation, but we may disagree about the meaning of "research." Dennett limits research to scientific investigations studying religious activities and organizations as social phenomena. In my opinion, such research, looking at religion from the outside, can be helpful but will never throw much light on the central mystery. The central mystery is the perennial sprouting of religious practices and beliefs in all human societies from ancient times until today. My mother, who was a skeptical Christian like me, used to say, "You can throw religion out of the door, but it will always come back through the window." I recently experienced a vivid demonstration of the truth of my mother's words. I went with my wife to visit the monastery of

Sergiev Posad north of Moscow, the ancient headquarters of the Russian Orthodox Church. The young guide who showed us around said almost nothing about the ancient buildings and works of art that we were supposed to be admiring. Instead she talked for an hour about her own faith and the mystical influences that she felt emanating from the old saints of the church in their tombs. After three generations of atheistic government and official suppression of religion, here it was sprouting again from its roots.

Let me state frankly my own philosophical prejudices in opposition to Dennett. As human beings, we are groping for knowledge and understanding of the strange universe into which we are born. We have many ways of understanding, of which science is only one. Our thought processes are only partially based on logic, and are inextricably mixed with emotions and desires and social interactions. We cannot live as isolated intelligences, but only as members of a working community. Our ways of understanding have been collective, beginning with the stories that we told one another around the fire when we lived in caves. Our ways today are still collective, including literature, history, art, music, religion, and science. Science is a particular bunch of tools that have been conspicuously successful for understanding and manipulating the material universe. Religion is another bunch of tools, giving us hints of a mental or spiritual universe that transcends the material universe. To understand religion, it is necessary to explore it from the inside, as William James explored it in *The Varieties of Religious Experience*. The testimony of saints and mystics, including the young lady at Sergiev Posad, is the raw material out of which a deeper understanding of religion may grow.

The sacred writings, the Bhagavad Gita and the Koran and the Bible, tell us more about the essence of religion than any scientific study of religious organizations. The research that Dennett advocates, using only the scientific tool kit that was designed for a different purpose, will always miss the goal. We can all agree that religion is a natural phenomenon, but nature may include many more things than we can grasp with the methods of science.

The best source of information about modern Islamic terrorists that I know of is a book, *Understanding Terror Networks*, by Marc Sageman.[2] Sageman is a former United States foreign service officer who worked with the Mujahideen in Afghanistan and Pakistan. In Chapter 5 of his book, he describes in detail the network that planned and carried out the September 2001 attacks on the United States. He finds that the bonds holding the group together, during its formative years in Hamburg, were more personal than political. He concludes: "Despite the popular accounts of the 9/11 perpetrators in the press, in-group love rather than out-group hate seems a better explanation for their behavior."

To end this review, I would like to introduce another recently published book, *Kamikaze Diaries: Reflections of Japanese Student Soldiers*, by Emiko Ohnuki-Tierney.[3] This contains extensive extracts from diaries written by seven of the young men who died in suicidal missions or as kamikaze pilots in the closing months of World War II. The diaries give us firsthand testimony of the thoughts and feelings of these young soldiers who knew that they were fated to die. Their thoughts and feelings are astonishingly lucid and free from illusions. Some of them expressed their feelings in poetry. All of them were highly educated and familiar with Western literature in several languages, having spent most of their brief lives in reading and writing. Only one of them, Hayashi Ichizo, was religious, having grown up in a Japanese Christian family. His Christian faith did not make self-sacrifice easier for him than for the others. He had read Kierkegaard's *Sickness unto Death* and carried it with him on his final mission together with his Bible.

All of the young men, including Hayashi, had a profoundly tragic

---

2. University of Pennsylvania Press, 2004.

3. University of Chicago Press, 2006.

view of life, mitigated only by happy memories of childhood with family and friends. They were as far as it was possible to be from the brainwashed zombies that contemporary Americans imagined to be piloting the kamikaze planes. They were thoughtful and sensitive young men, neither religious nor nationalistic fanatics.

Here I have space to mention only one of them, Nakao Takanori, who must speak for the rest. Nakao left a poem beginning, "How lonely is the sound of the clock in the darkness of the night." In his last letter to his parents, a week before his death, he wrote,

> At the farewell party, people gave me encouragement. I did my best to encourage myself. My co-pilot is Uno Shigeru, a handsome boy, aged nineteen, a naval petty officer second class. His home is in Hyogo Prefecture. He thinks of me as his elder brother, and I think of him as my younger brother. Working as one heart, we will plunge into an enemy vessel. Although I did not do much in my life, I am content that I fulfilled my wish to live a pure life, leaving nothing ugly behind me.

We have no firsthand testimony from the young men who carried out the September 11 attacks. They were not as highly educated and as thoughtful as the kamikaze pilots, and they were more influenced by religion. But there is strong evidence that they were not brainwashed zombies. They were soldiers enlisted in a secret brotherhood that gave meaning and purpose to their lives, working together in a brilliantly executed operation against the strongest power in the world. According to Sageman, they were motivated like the kamikaze pilots, more by loyalty to their comrades than by hatred of the enemy. Once the operation had been conceived and ordered, it would have been unthinkable and shameful not to carry it out.

Even after recognizing the great differences between the circumstances of 1945 and 2001, I believe that the kamikaze diaries give us our best insight into the state of mind of the young men who caused us such grievous harm in 2001. If we wish to understand the phenomenon of terrorism in the modern world, and if we wish to take effective measures to lessen its attraction to idealistic young people, the first and most necessary step is to understand our enemies. We must give respect to our enemies, as courageous and capable soldiers enlisted in an evil cause, before we can understand them. The kamikaze diaries give us a basis on which to build both respect and understanding.

V

Bibliographical Notes

1. This essay, which appeared in *The New York Review of Books*, May 25, 1995, was originally a lecture given at a conference in Cambridge, England, in November 1992. It was published in the proceedings of the conference, *Nature's Imagination: The Frontiers of Scientific Vision*, edited by John Cornwell (Oxford University Press, 1995).

2. This essay, which appeared in *The New York Review of Books*, April 10, 1997, is an abbreviated version of a lecture given at the Hebrew University of Jerusalem in May 1995. The lecture was published as Chapter 5 in Freeman Dyson, *Imagined Worlds* (Harvard University Press, 1997).

3. Foreword to Thomas Gold, *The Deep Hot Biosphere* (Springer-Verlag, 1999).

4. Review of Michael Crichton, *Prey* (HarperCollins, 2002), in *The New York Review of Books*, February 13, 2003.

5. Review of Vaclav Smil, *The Earth's Biosphere: Evolution, Dynamics, and Change* (MIT Press, 2002), in *The New York Review of Books*, May 15, 2003.

6. Review of Thomas Levenson, *Einstein in Berlin* (Random House, 2003), in *Nature*, April 24, 2003.

7. Review of Tom Stonier, *Nuclear Disaster* (Meridian Books, 1963), published in *Disarmament and Arms Control* (1964), pp. 459–461.

8. "Generals," from Chapter 13 in Freeman Dyson, *Weapons and Hope* (Harper and Row, 1984).

9. "Russians," from Chapter 15 in Dyson, *Weapons and Hope*.

10. "Pacifists," from Chapter 16 in Dyson, *Weapons and Hope*.

11. This essay, which appeared in *The New York Review of Books*,

March 6, 1997, is another piece of the same lecture from which Chapter 2 was taken, also published in Chapter 5 of Dyson, *Imagined Worlds.*

12. Preface to *Ending War: The Force of Reason: Essays in Honor of Joseph Rotblat*, edited by Maxwell Bruce and Tom Milne (Palgrave Macmillan, 1999). Reproduced with permission of Palgrave Macmillan.

13. Review of Max Hastings, *Armageddon: The Battle for Germany, 1944–1945* (Knopf, 2004); and Hans Erich Nossack, *The End: Hamburg, 1943*, translated from the German and with a foreword by Joel Agee and with photographs by Erich Andres (University of Chicago Press, 2004), in *The New York Review of Books*, April 28, 2005.

14. Review of Yuri Manin, *Mathematics and Physics*, translated from the Russian by Ann and Neil Koblitz (Birkhäuser, 1981); and Paul Forman, *Weimar Culture, Causality, and Quantum Theory, 1918–1927: Adaptation by German Physicists and Mathematicians to a Hostile Intellectual Environment*, Vol. 3 of *Historical Studies in the Physical Sciences* (University of Pennsylvania Press, 1971), published in *Mathematical Intelligencer*, Vol. 5 (1983), pp. 54–57.

15. Review of Edward Teller with Judith Shoolery, *Memoirs: A Twentieth-Century Journey in Science and Politics* (Perseus, 2001), published in *American Journal of Physics*, Vol. 70 (2002), pp. 462–463.

16. Review of Timothy Ferris, *Seeing in the Dark: How Backyard Stargazers Are Probing Deep Space and Guarding Earth from Interplanetary Peril* (Simon and Schuster, 2002), in *The New York Review of Books*, December 5, 2002.

17. Review of James Gleick, *Isaac Newton* (Pantheon, 2003), in *The New York Review of Books*, July 3, 2003.

18. Review of Peter Galison, *Einstein's Clocks, Poincaré's Maps: Empires of Time* (Norton, 2003), in *The New York Review of Books*, November 6, 2003.

19. Review of Brian Greene, *The Fabric of the Cosmos: Space, Time, and the Texture of Reality* (Knopf, 2004), in *The New York Review of Books*, May 13, 2004.

20. This chapter is the text of a talk given at the Institute for Advanced Study in Princeton on October 27, 2004, to celebrate Oppenheimer's hundredth birthday. The extract from Lansing Hammond's letter of 1979 was previously published in a preface that I wrote for *Atom and Void*, a collection of Oppenheimer's public lectures (Princeton University Press, 1989). Other parts of the chapter are borrowed from Chapter 11, "Scientists and Poets," of my book *Weapons and Hope*.

21. Review of Brian Cathcart, *The Fly in the Cathedral: How a Group of Cambridge Scientists Won the International Race to Split the Atom* (Farrar, Straus and Giroux, 2004); and Alan Lightman, *A Sense of the Mysterious* (Pantheon, 2005), in *The New York Review of Books*, February 24, 2005.

22. Review of Flo Conway and Jim Siegelman, *Dark Hero of the Information Age: In Search of Norbert Wiener, the Father of Cybernetics* (Basic Books, 2005), in *The New York Review of Books*, July 14, 2005.

23. Review of Richard Feynman, *Perfectly Reasonable Deviations from the Beaten Track: The Letters of Richard P. Feynman*, edited and with an introduction by Michelle Feynman (Basic Books, 2005), in *The New York Review of Books*, October 20, 2005.

24. Bernal Lecture given at Birkbeck College, London, May 1972, published as Appendix D, pp. 371–389, to *Communication with*

Extraterrestrial Intelligence, edited by Carl Sagan (MIT Press, 1973).

25. Review of Richard Feynman, *The Meaning of it All: Thoughts of a Citizen-Scientist* (Addison-Wesley, 1998); and John Polkinghorne, *Belief in God in an Age of Science* (Yale University Press, 1998), in *The New York Review of Books*, May 28, 1998.

26. Foreword to *The Pleasure of Finding Things Out: The Best Short Works of Richard Feynman*, edited by Jeffrey Robbins (Perseus, 1999). Copyright © 1999 by Michelle Feynman and Carl Feynman. Reprinted by permission of Basic Books, a member of Perseus Books, LLC.

27. Review of Georges Charpak and Henri Broch, *Debunked! ESP, Telekinesis, and Other Pseudoscience*, translated from the French by Bart K. Holland (Johns Hopkins University Press, 2004), in *The New York Review of Books*, March 24, 2004.

28. Preface to Olaf Stapledon, *Star Maker*, edited and with an introduction by Pat McCarthy (Wesleyan University Press, 2004).

29. Review of Daniel C. Dennett, *Breaking the Spell: Religion as a Natural Phenomenon* (Viking Penguin, 2006) in *The New York Review of Books*, June 22, 2006.

About the Author

FREEMAN DYSON has spent most of his life as a professor of physics at the Institute for Advanced Study in Princeton, taking time off to advise the US government and write books for the general public. He was born in England and worked as a civilian scientist for the Royal Air Force during World War II. He came to Cornell University as a graduate student in 1947 and worked with Hans Bethe and Richard Feynman, producing a user-friendly version of the theory of atoms and radiation. He also worked on nuclear reactors, solid-state physics, ferromagnetism, astrophysics, and biology, looking for problems where elegant mathematics could be usefully applied.

Dyson's books include *Disturbing the Universe* (1979), *Weapons and Hope* (1984), *Infinite in All Directions* (1988), *Origins of Life* (1986, second edition 1999), *The Sun, the Genome and the Internet* (1999), and *A Many Colored Glass: Reflections on the Place of Life in the Universe* (2007). He is a fellow of the American Physical Society, a member of the National Academy of Sciences, and a fellow of the Royal Society of London. In 2000 he was awarded the Templeton Prize for Progress in Religion.